The Waltham Book of Companion Animal Nutrition

*formerly The Waltham Book
of Dog & Cat Nutrition*

Edited by

I. H. BURGER

*Waltham Centre for Pet Nutrition
Melton Mowbray, Leicestershire*

PERGAMON PRESS

OXFORD · NEW YORK · SEOUL · TOKYO

U.K.	Pergamon Press Ltd, Headington Hill Hall, Oxford OX3 0BW, England
U.S.A.	Pergamon Press Inc, 660 White Plains Road, Tarrytown, New York 10591-5153, U.S.A.
KOREA	Pergamon Press Korea, KPO Box 315, Seoul 110-603, Korea
JAPAN	Pergamon Press Japan, Tsunashima Building Annex, 3-20-12 Yushima, Bunkyo-ku, Tokyo 113, Japan

First edition 1993

Library of Congress Cataloging in Publication Data
A catalogue record for this book is available from the Library of Congress

British Library Cataloguing in Publication Data
A catalogue record for this book is available from the British Library

ISBN 0 08 040843 5 Hardcover
ISBN 0 08 040844 3 Flexicover

DISCLAIMER

Whilst every effort is made by the Publishers to see that no inaccurate or misleading data, opinion or statement appear in this book, they wish to make it clear that the data and opinions appearing in the articles herein are the sole responsibility of the contributor concerned. Accordingly, the Publishers and their employees, officers and agents accept no responsibility or liability whatsover for the consequences of any such inaccurate or misleading data, opinion or statement.

Printed in Great Britain by BPCC Wheatons Limited, Exeter

Contents

Preface

It was twenty years ago when I first entered the field of companion animal nutrition, which was then not as well documented as, for example, human or farm animal nutrition. Nevertheless, the fascination of the subject was, and still is, its unique position between these other two disciplines and the challenge it proposes. In essence, it is to design a single dietary regimen (often a single product) which satisfies the nutritional needs of the animal and with the overall aim of a long and healthy life. Pets are important for people (a subject equally fascinating but outside the scope of this book) and it is vital that they are fed properly. I have been privileged to be involved in the rapid expansion of companion animal nutrition and it is gratifying to look back over the advances that have been made, both in the detail and status of the subject. These advances are reflected in the parent (1988) and grandparent (1982) versions of this volume which were produced under the distinguished editorship of Andrew Edney and I am particularly indebted to him for his eloquent foreword to this edition.

I must also thank Caroline Franklin and Fenella Jones of the Grayling Company for reviewing the script, and Pergamon Press (in particular Marion Jowett and Mandy Sketch) for invaluable assistance with the project and considerable patience with the editor. Finally, I give special thanks to my colleagues at Waltham for much help and encouragement, in particular, of course, my fellow authors for producing contributions which were a pleasure to edit.

Ivan Burger
April 1993

List of Contributors

DEREK BOOLES:

Derek completed a student apprenticeship in industrial chemistry at Fisons covering various aspects of research, product development and sales development. By part-time study he gained an HNC in chemistry. In 1967 he joined Pedigree Petfoods working in the Organic Research laboratory. He then became an Associate of the Institute of Biology and in 1969 moved to the Waltham Centre for Pet Nutrition (WCPN) where he has worked on many aspects of the nutrition of the dog and cat.

IVAN H. BURGER BSc PhD:

Ivan graduated in physiology and biochemistry from Southampton University in 1968. After graduation he joined the Leatherhead Food Research Association to carry out research into the biochemistry of meat curing, with particular reference to the involvement of the mitochondrial electron transport chain. This work was presented as a PhD thesis (awarded 1972) on a collaborative basis with the Biochemistry Department of Surrey University. In 1973 Ivan moved to WCPN. His first job was Nutritional Biochemist and he conducted work on the nutritional requirements of the cat, in particular protein and amino acids. Following this, he assumed responsibility for food safety in relation to additives and contaminants while retaining an interest in the nutritional work. In 1987 he started a two year assignment into the interactions between people and companion animals with particular reference to the effects of pet animals on the health of their owners. In 1989 Ivan rejoined the nutrition group as Senior Nutritionist in charge of studies in the dog, while retaining responsibilities for food safety. In 1992 he further extended his job to include responsibility for equine nutrition.

KAY E. EARLE BSc PhD:

Kay graduated in biochemistry from Manchester Polytechnic in 1981. She obtained a PhD in bio-organic chemistry from Imperial College, University of London in 1985. She joined the

Department of Medicine, University of Cambridge in 1985 as a post-doctoral research fellow. Her work included investigations of the energy and trace mineral metabolism of lean and obese individuals with and without non-insulin dependent diabetes. She joined WCPN in 1988 as the cat nutritionist and expanded her responsibilities into the areas of bird and fish nutrition following her promotion to Senior Nutritionist in 1989.

JANEL V. JOHNSON BSc PhD:

In 1983 Janel graduated from the University of Nottingham with an honours degree in Animal Physiology and Nutrition. She was awarded a PhD for her work on blood pressure regulation at the Medical School, Nottingham University in 1986. She spent a further two years at Nottingham working as a post-doctoral fellow on disorders of the spinal cord. She joined WCPN in 1988 and was promoted to her present position of Nutritionist in 1991. She is currently in charge of digestibility studies in dogs. Janel is involved with dog training and has trained her own Obedience Champion.

VÉRONIQUE LEGRAND-DEFRETIN DEA Doctorat:

Véronique graduated in biochemistry from Paris VII University in 1982. She subsequently obtained diplomas in food science and nutrition. She then joined the French National Institute for Agricultural Research to carry out research into the physiology of bile secretion in pigs, with particular reference to the enterohepatic circulation of bile salts (Thèse de Doctorat, awarded in 1988). For her post-doctoral research, she studied the treatment of gallstones at the National Institute for Scientific Research, and subsequently on biliary secretion on liver transplant patients at the National Institute for Health and Medical Research. Véronique joined WCPN in 1991 as a Nutritionist and is currently in charge of lifestage and energy studies in dogs.

IAN E. MASKELL BSc PhD:

Ian graduated from the department of Animal Physiology and Nutrition, Leeds University in 1986. He was awarded a PhD in 1990 by Newcastle University for studies of nutritional and toxicological aspects of rapeseed meal in porcine diets. Ian joined WCPN in 1990 and is currently involved in the research development of clinical diets, with an emphasis on gastro-intestinal-related illness.

HELEN S. MUNDAY BSc PhD:

Helen graduated in Agricultural (Animal) Science from Leeds University in 1986. She was awarded her MSc for research in meat science from Nottingham University in 1987. Helen joined Pedigree Petfoods in 1987 in the Materials Development Department, initially working on the control of raw material quality and then moved to the Waltham Centre for Pet Nutrition in 1989 in the role of Project Scientist. In 1991 Helen was promoted to her current position of Nutritionist in which she has responsibility for research into the optimum feeding of cats through the lifestages.

HELEN M. R. NOTT BSc PhD PCertMg:

Helen studied Biology at York University from 1982–1985 specialising in Ecology and Animal Behaviour. In 1985 she went to Reading University where she completed her PhD on the behaviour of wild rats. During this time she also acted as consultant for a rodent control project in West Africa. In 1988 she joined WCPN in the behaviour team working on the feeding behaviour of cats and dogs. In 1991 she was promoted to Bird Nutritionist and in 1992 her role expanded to include work on cat nutrition. Since 1991 she has been the scientific editor of the *Journal of Feline Advisory Bureau.*

MARINUS C. PANNEVIS MSc:

Marinus Pannevis graduated *cum laude* in 1988 at the University of Wageningen, the Netherlands, in fish nutrition and fish immunology. From 1989 he spent two years at the Zoology Department of the University of Aberdeen in Scotland researching the energetic cost of protein synthesis of fish. Since 1990 he has been Head of the Waltham Aquacentre and as fish nutritionist, responsible for the AQUARIAN® ornamental fish food development.

PHILIP M. SMITH:

Phil studied part-time at Leeds College of Technology for an HNC in Chemistry whilst working as a laboratory assistant and pilot plant technician from 1955–1964 for J. E. Sturge Ltd. In 1964 he joined Pedigree Petfoods in the Quality Information Department, moving to the Nutrition Laboratory (now WCPN) in 1968, working in areas of palatability, behaviour and nutrition of companion animals. He has now accumulated 25 years of experience in this environment and is currently a Project Scientist in areas of cat nutrition.

E. JEAN TAYLOR BSc MSc:

Jean graduated from Newcastle Polytechnic in 1986 with a degree in Environmental Science. In 1988 she obtained a Masters Degree in Animal Nutrition from Newcastle University having specialised in the area of carbohydrate digestibility. This was followed by three years at Nottingham University studying for a PhD on the subject of starch digestibility. Jean joined WCPN in 1991 in her current position as Project Scientist in cat and bird nutrition.

WILFRIED E. TIEGS Dr. Med. Vet.:

Wilfried Tiegs graduated as a veterinarian from the Hanover Veterinary School in 1985. He obtained his PhD for studies on the use of drug therapy in pigs with respiratory tract disease from the Animal Health Institute in Hanover in 1986. For four years he worked as a veterinary surgeon, especially in equine practices. In 1990 Wilfried joined Waltham to head the Waltham Centre for Equine Nutrition and Care. Since July 1992 he has been working as a consultant and, from 1993 onwards, has run his own veterinary practice.

Foreword

The *Waltham Book of Dog and Cat Nutrition* appeared in 1982 and a second edition was published six years later. Now, after a similar period a new work has become available. The title is not the only difference between this book and its predecessors. The *Waltham Book of Companion Animal Nutrition* now embraces birds, ornamental fish and horses. All such species now represent a significant percentage of modern day companion animals.

This book reflects the greatly increased body of knowledge in the whole area. However, the objectives remain as before. That is, to present the best information on the subject in an accessible, readable way. The readership remains as it was, that is student veterinarians and nurses, as well as those breeders and owners of companion animals who take an active interest in their feeding. However, the information they need to know is very much greater than it was even a few years ago.

In the 1990s the task of making nutrition comprehensive is very much more demanding than it was in the past. However, the team at the Waltham Centre is unmatched in the world. All of them have done a magnificent job making this publication the best yet in the field.

Now that the science of nutrition has moved much more into the professional and general consciousness, it is of the utmost importance that a standard work is available which crystallises all the key points on the subject. The *Waltham Book of Companion Animal Nutrition* does this to the highest standard.

Andrew Edney

CHAPTER 1

Foods Fit for Companion Animals

DEREK BOOLES

Introduction

Animals feed in order to obtain all the nutrients and the energy necessary to sustain a healthy life and for successful procreation. Feeding any pet or companion animal should also be an occasion enjoyed by both the animal and the owner. This can only be achieved when suitable foods are offered which are readily eaten by the animal and which provide it with a balanced diet. One of the major preoccupations for the owner of a pet or companion animal is keeping it fit and healthy and a major part of this is feeding the correct diet.

Any consideration of what constituents make up a 'proper' diet for an animal must take into account both its lifestyle and its lifestage. Most pet owners keep their animals essentially as companions but some, especially dog and horse owners, expect them to perform other duties, for example, to hunt, race, keep down vermin, pull loads or for herding. These activities may demand different types of diet and therefore the food ingredients and the proportions from which they are composed will also differ. Individuals may be simply at maintenance, growing rapidly, gestating, feeding young, or just growing old! Each lifestage makes its own particular demands

on a diet, thus a young growing animal has different nutritional needs to an adult, or a pregnant or a lactating one. Species which have a large size range such as dogs and horses may have special requirements, not least influenced by the variation of food allowance with body size. Each of these lifestages, or an animal with a particular lifestyle, needs a diet which is balanced in order to ensure that it remains fit and healthy. A balanced diet can be defined as that mixture of ingredients which provide all the energy and essential nutrients needed to maintain the animal in health appropriate to its lifestyle and lifestage. The question then, is to determine what foods can be used to provide pets with a balanced diet.

Evaluation

To any owner of a pet the most rewarding service that may be performed is to keep their companion healthy and fit, so that a long and happy life can be enjoyed. Probably the single most important aspect of achieving those ends lies with the diet that is provided. In many ways the nutritional feats expected of pet foods often exceed those of both human food and farm animal feed. For example, humans by and large have free access to a wide variety of foods, whereas pets are usually restricted to what their owners give them or allow them access to. Farm animals are fed for quite different reasons to pets: they are rarely expected to live out their natural lives and their feed is designed for efficient production of weight gain, milk, eggs and so on. For pets the main aim is the same as for their owners, that is a long, happy and healthy life.

The animals discussed in this book—dogs, cats, birds, fish and horses—probably represent the most popular pets worldwide and some of the longest lived. Horses can live up to 25 years, cats up to 20 years and dogs up to 15. Some fish and birds, separated from their natural predators, may live just as long as their four-legged cousins. This reinforces the long healthy life objective required of the feeding regime. However, the five groups of pets actually represent a very wide spread of species type. All belong to the phylum chordata but horses, dogs and cats are in the class mammalia, birds aves and fish are often collectively described as pisces, although they are more correctly sub-divided into three classes. Dogs, cats and horses are all mammals and cover an extremely wide range of body weight: an adult cat is about 4–5 kg whereas a large horse can weigh up to 700 kg. In fact dogs and horses are unique among mammals in displaying such a wide range of body weight within a *single* species. A survey of adult dogs conducted by the Waltham Centre for Pet Nutrition (WCPN) found that at 1.1 kg the Chihuahua was the lightest and at 115 kg the St Bernard was the heaviest (Johnson, Burger and Markwell, in press). Similarly the Miniature Pigmy horse can weigh as little as 50 kg while a Shire horse can weigh up to 1000 kg. Birds kept as pets again cover a very wide range of types. They lay eggs rather than give birth to live young (oviparous) and their rate of metabolism is generally much higher than mammals. With fish we have entered the aquatic medium and this obviously poses new challenges in terms of feeding. Fish are cold-blooded animals (poikilotherms) and allow their body temperature, and therefore metabolic rate, to be

controlled by the temperature of the surroundings. In contrast the other species are all warm-blooded (homoiotherms) and maintain their body temperature at around 35–40°C which is usually well above that of the surroundings.

For any species the only effective method for evaluating and validating a diet is to feed it to the target animal and to make a series of objective measurements relative to biological performance. Diets may be arrived at by trial and error (i.e. experience) but scientific guarantees of nutritional adequacy demand resources usually only available to professional nutritionists. Leading manufacturers of prepared pet foods maintain facilities for carrying out complex and demanding nutritional assessments of their products. From such manufacturers the wide variety of products available may be used confidently and safely in the knowledge that their quality and nutritional status are assured.

The basis of any validation of the nutritional adequacy of foods for pets must embrace a knowledge of the animal's requirements for specific nutrients and to match those against the nutrient content of the food in question and its biological performance. Additionally it is important to ensure that the food is safe, i.e. contains no toxic elements and that it is palatable. It is worth re-stating that uneaten food is nutritionally worthless, whatever its nutrient content.

Palatability

The palatability of food is a complex subject including a knowledge of the factors affecting appetite and behaviour, as well as an understanding of the taste, smell and texture of food and their inter-relationships. A book on nutrition cannot adequately deal with palatability but the importance of a highly palatable food cannot be over emphasised. Making products which are consistently well eaten over extended periods requires a great deal of expertise and experience. Reputable manufacturers of prepared pet foods have developed objective measures for assessing palatability in order to ensure that any given recipe offers the consistent level of palatability expected by the animal and the owner.

Individual animals, just like humans, have sharply individual preferences. To accommodate these tastes, whether the food is canned, dry or in the form of seeds or flakes, large numbers of individual animals are studied for their likes and dislikes. In this way recipes can be developed which consistently give enjoyment to both the pet and the owner. These tests, which basically study intake under defined conditions, give rise to a large amount of complex statistical data, the interpretation of which gives a measure of relative palatability in terms of preference and acceptance.

Pet animals enjoy good quality food and find low quality foods less attractive. This may influence intake and result in 'problem' feeders, or even give rise to nutritional problems associated with poor intake. In particular, most dogs and cats enjoy change and so benefit from a variety in their choice of food. However, sudden changes of food type may produce digestive upsets. This is minimised by many manufacturers who produce variety within brands, such that the novelty of a new taste may be enjoyed without a disturbance of digestion.

Assessing Performance

For the manufacturer it is self evident that it is in their own best interests to ensure that their products satisfy their clients and this effectively covers *two* customers—the owner and the pet itself. Foods must conform to the owners' idea of suitability as well as the pets' own nutritional requirements and palate. There are various areas of legislation which govern nutritional performance of pet foods and these are summarised in Appendix III. In addition the pet food industry sets various voluntary codes of practice. Standard tests are well defined for dogs and cats but these may have to be modified or special tests designed to perform similar evaluations for fish, birds and horses. The testing and labelling of food require certain criteria to be carefully

defined. These mainly refer to the lifestage of the animal and whether the food is complete. The main lifestages are as follows:

- growth (young independent of mother)
- maintenance
- reproduction
- rearing of young e.g. lactation in the mammal.

Thus a food designated as 'complete for growth in dogs' will have to supply a complete diet for puppies without the necessity for *any other* food or supplements. The only additional requirement is for water. (The exception to this rule are foods which are described as complementary. To become complete these foods need to be mixed with another food.) Similar criteria apply to the other lifestages.

However, there are other lifestyles within adult maintenance which may also require special nutrition. Two examples are working and elderly (senior) animals. The former is a particularly important aspect for the horse and dog. It is still unclear precisely what are the special needs for animals in these two categories but there is some evidence that certain nutrients need to be present at different concentrations compared with a 'normal' healthy adult. These particular aspects will be discussed further in Chapters 5 and 8.

Millions of companion animals obtain a substantial proportion of their daily nutrient intake from food prepared by pet food manufacturers. There is therefore an increasing obligation on the manufacturer (reinforced by international controls and regulations) to 'get it right' in providing the correct amounts and balance of nutrients to sustain a healthy life for companion animals. In a very competitive industry the search for new raw materials, processes and product forms to improve value for money cannot be allowed to compromise the nutritional package offered to the pet. Continued nutritional research to support and advance the current knowledge of companion animal nutrition described in the following chapters is essential to take advantage of fresh opportunities and to pre-empt potential challenges that new technology and the demands of the business may generate.

CHAPTER 2

A Basic Guide to Nutrient Requirements

IVAN H. BURGER

Introduction

Like all other living creatures, companion animals require food to stay alive and healthy. Food may be defined as 'any substance which is capable of nourishing the living being'. A more complete description is that food is any solid or liquid which, when ingested, can supply any or all of the following:

(a) energy-giving materials from which the body can produce movement, heat or other forms of energy;

(b) materials for growth, repair or reproduction;

(c) substances necessary to initiate or regulate the processes involved in the first two categories.

The components of food which have these functions are called nutrients and the foods or food mixtures which are actually eaten are referred to as the diet. The main types of nutrients present in foods are:

Carbohydrates—these provide the body with energy and may also be converted into body fat. This group includes simple sugars (such as glucose) and larger molecules (such as starch) which consist of chains of the simpler sugars.

Fats—these provide energy in the most concentrated form, releasing about double the amount of energy per unit weight than either carbohydrates or protein. Fats aid in the absorption of the fat-soluble vitamins and supply types of fat usually referred to as the essential fatty acids (EFA). These, as their name suggests, are required for certain important body functions and are as important as individual vitamins or minerals. The EFA will be discussed in more detail later in this chapter.

Proteins—these are important because they provide amino acids which are involved in the growth and repair of body tissue. The component amino acids can also be metabolised to provide energy, about the same amount per unit weight as carbohydrates.

Minerals and trace elements—the 'major minerals' are substances like calcium and phosphorus which are used in growth and repair and make up most of the skeletal and tooth structure. Substances required in smaller quantities such as iron, copper and zinc are usually referred to as trace elements.

Vitamins—these help to regulate the body processes and are usually considered as two categories, the fat-soluble and water-soluble groups. In the former are vitamins A, D, E and K; the latter group includes vitamins of the B complex (such as thiamin) and vitamin C.

The other important constituent of food is water and although this is not generally regarded as a nutrient, it is essential to life. Water balance is discussed in greater detail in the next chapter. The need for water is second only to the need for oxygen, the other vital element not included in the list above.

Hardly any foods contain only one nutrient: most are complex mixtures which consist of a variety of carbohydrates, fats and proteins together with water. Minerals and vitamins (especially the latter) are usually present in much smaller amounts.

An Adequate Intake

An adequate intake of nutrients is essential for the health and activity of the animal, but how much is adequate? Compared with the requirements of the adult animal, there are additional needs for the more demanding stages of the lifecycle such as growth, pregnancy and lactation. In the case of companion animals, it is often possible to investigate their needs for nutrients and to obtain more precise values than is possible for man. The minimum quantity of an individual nutrient which must be supplied each day for proper body metabolism is usually referred to as the minimum daily requirement (MDR). It must be remembered, however, that by definition MDR values are *minima* and appropriate allowances must be made for individual variation, physical activity, breed, weight, sex and stage of development. Furthermore, there are other factors which may be taken into account, in particular the availability of nutrients in foods and these will be discussed later in this chapter. In view of all these considerations, it is more practical to use the MDR data to derive values for a recommended daily allowance (RDA) as a preferred guide for nutritional adequacy. For any animal the RDA is designed to ensure that the needs of virtually all the normal healthy individuals in the population are covered. It follows that the RDA will always be in excess of MDR (except

for energy which is discussed below) and experimentally determined requirements will be less than recommended intakes. It also follows that a diet may contribute *less* than the RDA, but still provide an adequate nutrient intake for a certain proportion of the population.

An equally important aspect is the application of RDA (or MDR) to a food or mixture of foods i.e. the diet. Requirements will initially be assessed as a quantity of nutrient ingested by the animal and will usually have units of intake per unit body weight per day. But ultimately the most useful and relevant way to express this value is a concentration in the diet. This raises the question of the quantity of different types of foods eaten by different animals. Foods have different compositions (from canned to dry) and animals, particularly dogs and horses, show a wide variation in size from breed to breed. The link between these variables is energy.

Energy

Content in the diet

Energy in food differs from the nutrients, in that its intake must be kept close to requirements. Intakes in excess of requirements are undesirable and eventually lead to obesity. The energy content of a diet is derived from carbohydrates, fats and protein and the amount of each of these nutrients in a food will determine its energy content. Water has no energy value, so the energy density of food varies inversely with its moisture content. In nutrition, energy has usually been expressed in terms of kilocalories (kcal) where 1 kcal is defined as the quantity of heat required to raise the temperature of 1 kg of water by 1 degree Celsius. A more recent convention is to express energy in terms of the joule which is more difficult to define in familiar terms and is based on a mechanical or electrical equivalent of heat (one watt is a joule per second). The conversion between the two systems is such that 1 kcal is equivalent to about

4.2 kilojoules (kJ), or 0.0042 megajoules (MJ) i.e. 1 MJ = 1000 kJ.

The body obtains energy by oxidising ('burning') food but, unlike the burning process in a boiler or engine, the energy is released gradually by a series of complex chemical reactions, each regulated by an enzyme. Enzymes are special proteins which control the rate of chemical reactions and more importantly, enable these complex changes to take place in the relatively mild conditions of the body. To bring about the same changes in a typical industrial process would require much more extreme conditions of temperature and pH or highly reactive ingredients. Many enzymes require the presence of vitamins or minerals to function properly and this aspect will be discussed in more detail when these nutrients are considered.

No animal is able to utilise *all* the energy from its food. Energy intake is therefore considered at three different levels: gross energy (GE), digestible energy (DE) and metabolisable energy (ME). Gross energy is the total energy released by complete oxidation of the food and is usually measured by burning it in an atmosphere of pure oxygen in an instrument (calorimeter) which accurately measures the heat released on combustion. Although a substance may have a high GE content, it is of no use unless the animal is able to digest and absorb it. The amount which is digested and absorbed is known as DE and equals GE minus faecal losses. Some of the absorbed food may only be partially available to the tissues, the remainder being lost in the urine. The energy which is ultimately utilised by the tissues is known as ME and is calculated as DE minus urinary loss. The DE and ME contents of foods depend both upon their composition and upon the species which eats them. The digestive systems of the animals discussed in this book differ markedly (see Chapter 3) and it would be surprising if the same food delivered the same proportion of absorbed nutrients (i.e. nutritional value) in every species. Even two fairly similar animals, such as the dog and cat, show differences in digestibility values when fed the same food (Kendall *et al.*, 1982). This may be partly because the dog's digestive system is proportionately longer than the cat's and therefore is likely to be more efficient. In addition to these species differences, there will be variations between individual animals in their own metabolic efficiency. So the only way to obtain a meaningful measurement of the ME content of a food in a particular animal is to feed it to as large a group as possible and measure energy (using the calorimeter technique) in food, faeces and urine. This system, although perfectly feasible, is time consuming and costly and is not possible without access to specialised animal facilities. Therefore over the years simple formulae have been developed which give reasonable approximations of the ME in a food from its carbohydrate, fat and protein contents, allowing for the losses in absorption and efficiency. The factors originally used were those derived from studies in man, but more accurate figures for dogs and cats have been published which are compiled from feeding studies (NRC 1985, 1986)—see Appendix II. Studies conducted at the Waltham Centre for Pet Nutrition (WCPN) in which DE measured *in vivo* was compared with ME predicted from factors suggest that the new values generally give a good estimate of the energy available to the dog and cat from typical commercial pet foods. For the other animals, these food energy factors are less well defined and will be discussed in the appropriate chapters.

Requirement

Energy expenditure can be divided into two parts, basal metabolic rate (BMR) and thermogenesis. BMR is the amount of energy required to keep the body 'ticking over', that is it represents the energy needed to meet the cost of essential work done by the cells and the organs. This includes such processes as respiration, circulation and kidney function. Many factors determine BMR in any individual including body weight and composition, age and hormonal status (particularly the thyroid hormones). As these factors change, so does the rate of basal metabolism, although such changes tend to occur slowly over long periods.

Additional energy expenditure comes under the collective title of thermogenesis. This can be the cost of digesting, absorbing and utilising nutrients (sometimes called the 'thermic effect of food' or 'dietary induced thermogenesis'), of muscular work or exercise, of stress or of the maintenance of body temperature in a cold environment. The intake of certain drugs or hormones can also be the cause of thermogenesis. Thermogenesis is simply an increase in metabolic rate over the basal level. In contrast to BMR the degree of thermogenesis can vary widely and quickly and may cause large daily variations in energy output. Of the two components of total energy expenditure, thermogenesis is the part capable of rapid adaptive response to changes in the internal or external environment.

As was mentioned earlier, energy expenditure or requirements can ultimately be measured in terms of heat loss or heat production and it is this fact which leads on to a key concept in the calculation of energy requirements for different animals. Clearly it is most convenient to express energy needs in terms of body weight, which is a familiar and easily measurable quantity. However, heat loss varies not with body weight but with surface area, a quantity which is very difficult (even daunting) to assess accurately. Nevertheless, as surface area varies with the square of linear dimensions and body weight with the cube (i.e. volume), heat loss *should* vary with body weight (W) adjusted mathematically in line with these two factors i.e. $W^{2/3}$ or $W^{0.67}$. The body weight thus modified is often referred to as the **metabolic** body weight or body size. It should be possible to use this value to compare energy requirements of animals of widely different body weights and this is a crucial aspect, especially as regards dog and horse nutrition. These two animals are unique among mammals in displaying such a wide variation in body weight within a *single* species (see Chapter 1). However, metabolic body weight is also useful in comparing energy requirements between, for example, different species of birds and between warm-blooded and cold-blooded animals both of which are covered in this book under the

umberella of companion animals. A detailed discourse on this subject could fill several volumes and is out of place here. Suffice it to say that much research has been and is still being conducted on the most relevant factor (or factors) to use for this concept. At present it appears that values between 0.67 and 0.75 reflect the energy requirements of dogs (Burger and Johnson, 1991; Manner, 1991; Rainbird and Kienzle, 1990) and birds (Lasiewski and Dawson, 1967), but this is not so clear for the horse (Pagan and Hintz, 1986). Based on our own and other investigations, the current WCPN equations for the *average* requirements of dogs are as follows:

$$E = 125 \ W^{0.75} \text{ kcal/day}$$
$$\text{or}$$
$$\bar{E} = 523 \ W^{0.75} \text{ kJ/day}$$

where W is body weight in kg (Legrand-Defretin, 1993). These are often referred to as **allometric** equations i.e. relating size to a function, in this case energy requirement. The values obtained assume a moderate amount of activity for a dog in a temperate climate. Extremes of environmental temperature will obviously influence requirements markedly. It must be appreciated that, as the equation involves an exponential function, it is important to define the units of body weight (e.g. g or kg) as this will affect the final result. Assuming this is done, comparisons can be made across a wide spectrum of species and further details can be found in the chapters on birds and fish. Cats show a relatively narrow range of body weight (from around 2.5 to 6.5 kg) and it might be expected that energy requirements could be related directly to body weight with little error. Nevertheless even in this species a logarithmic relationship to body weight has been reported with larger cats displaying a lower energy requirement per unit of body weight than smaller ones (Earle and Smith, 1991a).

Balance

Energy balance is achieved by the matching of input and output over long periods. The principal mechanism of control is thought to

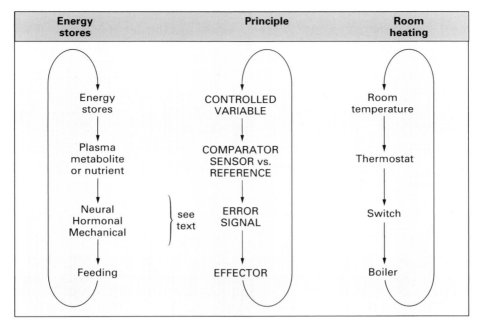

Energy stores	Principle	Room heating
Energy stores	CONTROLLED VARIABLE	Room temperature
Plasma metabolite or nutrient	COMPARATOR SENSOR vs. REFERENCE	Thermostat
Neural Hormonal Mechanical	ERROR SIGNAL	Switch
Feeding	EFFECTOR	Boiler

FIG 2.1: The negative feedback system.

be via regulation of intake although some variation of output is also thought to be important. Regulation of intake is considered to fit a negative feedback mechanism to a greater or lesser degree.

In its most straightforward form, negative feedback is a system where any change in the equilibrium of the system elicits a signal, provoking a response to oppose the initial change and correct the error. In the example (Fig. 2.1) the room thermostat is the *sensor* which detects any change in ambient temperature. The discrepancy between the room temperature and a set reference temperature is noted by the *comparator* which signals to the boiler or *effector*, switching it on or off. The heat output from the boiler restores the room temperature, which can be thought of as the *controlled variable* and causes the comparator to cease signalling to the boiler.

In this model of the regulation of energy balance, the controlled variable is the size of the energy stores. There are several feedback elements which may signal change, such as plasma nutrient and metabolite levels. Any discrepancies against set reference points indicate change in energy stores and stimulate

neural and hormonal activity which may initiate or inhibit feeding. The neural response involves 'feeding centres' in the brain which are probably not discrete 'hunger' and 'satiety' centres as once thought, but bundles of neurones covering several areas. Stimulation of these by electrodes can cause satiated animals to eat or prevent hungry animals from eating.

The hormonal response is more complex. Insulin stimulates feeding but it is not known whether this is a direct effect on the central nervous system or by causing peripheral hypoglycaemia (low glucose levels in the blood). Glucagon has the opposite effect to insulin, inhibiting feeding, as do oestrogens and luteinising hormone (female reproductive hormones). As all these hormones have roles other than the stimulation or inhibition of feeding, they cannot be the sole agents governing intake.

In addition to neural and hormonal mechanisms there are other, more direct stimuli to feeding. Contractions of an empty stomach are thought to cause the sensation of hunger and provoke feeding, whereas gastric distension inhibits eating.

This model provides a useful framework to the theories of regulation of energy intake. However the full explanation is still elusive and allowance must be made for some control of energy expenditure to regulate energy balance. For example, prolonged overfeeding results in an elevated metabolic rate which restricts the increase in energy stores. Nevertheless it cannot be overemphasised that it is *long-term* over- or underfeeding in companion animals (usually the former) which leads to problems. Even a small imbalance maintained for a long time will cause obesity (if the net difference is positive) or wasting (if the difference is negative). A 35 kg adult dog will require about 7500 kJ (1800 kcal) per day according to the WCPN equation. If this estimate is wrong by only 400 kJ/day (about 5%), then it could result in an increase of 2–3 kg in body weight if maintained for a year.

From the previous discussions it is clear that setting *precise* energy recommendations for any individual companion animal (especially the dog) is not feasible. The pet food manufacturer will use a good average estimate in compiling feeding guides but it is equally important for the pet owner to use these guidelines with discretion and to regulate food intake in relation to weight changes and the general appearance of the animal. This is especially necessary if the pet's own internal feedback system is not as exact as the theory might predict.

In any event, it is the energy requirement of the animal and the energy density of the food which determine the quantity of food eaten each day and thus the amount of each nutrient ingested. In this book nutrient concentrations are therefore usually expressed in terms of the dietary ME content so that the values are applicable to any type of food.

Nutrient Functions

In the following sections the precise functions of the various nutrients will be discussed. Much of the detail is common to all of the species considered in this book. Where there are special aspects, emphasis has generally been given in this chapter to the dog and cat. More specific details of birds, fish and horses will be found in Chapters 6, 7 and 8 respectively.

Carbohydrate

There is no known minimum dietary requirement for carbohydrate although the balance between carbohydrate and fat for exercise is of importance (see below). Based on investigations in the dog and with other species it is likely that most animals can be maintained without carbohydrates if the diet supplies enough protein from which the metabolic requirement for glucose can be derived. For example, it has been reported (Rosmos *et al.*,1981) that the consumption of a high-fat carbohydrate-free diet by bitches during gestation substantially reduced the survival of puppies compared with a control group receiving a diet containing 44% ME as carbohydrate. The effect was attributed to a severe hypoglycaemia in the former bitches at whelping. Nevertheless a study has been conducted at WCPN in which two diets were compared with regard to their ability to support Beagles and Labradors through pregnancy and lactation (Blaza *et al.*,1989). One diet supplied no available carbohydrate whereas the other contained 11% of ME as carbohydrate. No differences in performance were seen between these two treatments—both supported normal pregnancy and lactation. The contrast between the two studies can probably be explained by differences in dietary protein content. Romsos and co-workers used 26% protein energy whereas in the latter investigation the protein level was much higher (51% ME)—high enough to supply an adequate level of glucose. A similar effect was reported by Kienzle and Meyer (1989), who recommended a minimum protein content of 33% ME in a carbohydrate-free diet. Thus while carbohydrate is *physiologically* essential, it is not an indispensable component of the diet.

The carbohydrate source used in both of these studies was cooked starch and there is little doubt that this substance is readily digested. Individual disaccharides (i.e. containing two sugar units) such as sucrose (cane sugar) and lactose (milk sugar) are less well tolerated, especially the latter. The ability to metabolise these sugars is governed, respectively, by the amounts of the enzymes beta-fructofuronidase (sucrase) and beta-galactosidase (lactase), present in the intestine. Sucrase and lactase activities are certainly present in adult dogs and cats, although they are known to be higher in kittens and to decline with increasing age. If adult or young animals are suddenly given *large* amounts of lactose (for example a large bowl of milk) they may exhibit diarrhoea which is due partly to osmotic purgation and partly to bacterial fermentation (in the large intestine) of carbohydrate escaping digestion. Despite this, small quantities of these carbohydrates (say 5% of total energy) can be well tolerated by most animals although there will obviously be variations in the efficiency of individual animals in utilising these substances. Cats have generally less capacity than dogs to digest carbohydrate, due to lower enzyme activities in the small intestine (see Chapter 3).

Some work with dogs (Kronfeld *et al.*, 1977) has suggested that a carbohydrate-free, high-fat diet actually confers some advantages for prolonged strenuous running in racing sledge dogs, compared with diets containing up to 38% of ME as carbohydrate. These advantages included a higher oxygen-carrying capacity in the form of more red blood cells and haemoglobin. However, for normally active animals the inclusion of 40–50% of ME as dietary carbohydrate is unlikely to represent any disadvantage compared with a total fat and protein diet. In horses there is probably some requirement for carbohydrate, especially for work, but the precise needs are somewhat controversial (see Chapter 8).

Dietary Fibre

Dietary fibre refers to a group of carbohydrate-like substances such as cellulose, pectin and lignin. They are usually associated with plant materials and typically constitute the cell walls of plants. Dietary fibre sources include wholemeal cereals, root vegetables, fruit and gelling agents. These materials generally escape complete digestion in the small intestine and enter the large bowel largely unchanged. They have been the subject of much research, especially in man, where they are said to help prevent large bowel diseases such as diverticulosis, constipation and colon cancer. Nevertheless their role in the animal depends on many factors notably the physiology of the digestive system. In the dog and cat a *limited* amount of dietary fibre may provide bulk to the faeces and may therefore be useful in the prevention or management of constipation or diarrhoea under certain circumstances. In the horse, fibre probably has a more important role in the prevention of digestive problems. In addition the horse has the ability to ferment fibre via bacterial activity in its colon, to release a certain amount of usable energy (see Chapter 8).

Fat

Dietary fat serves as the most concentrated source of energy in the diet and lends palatability and an acceptable texture to foods. Like carbohydrates, fats are compounds of carbon, hydrogen and oxygen. Chemically, food fats consist mainly of mixtures of triglycerides, where each triglyceride is a combination of three fatty acids, joined by a unit of glycerol. The differences between one fat and another are largely the result of the different fatty acids in each. There are many different fatty acids found in foods and their chemical structures are characterised by the number of carbon atoms and double bonds. Saturated fatty acids have no double bonds, whereas the unsaturated variety have one or more; those containing more than one double bond are referred to as polyunsaturated. Most fats contain all of these types but in widely varying proportions.

It is difficult to give a precise requirement for total dietary fat. The only demonstrable need for fat is as a provider of EFA and a carrier of the fat-soluble vitamins. These functions will determine the requirement for fat together with the need to provide a certain content in the diet to achieve the necessary energy density and palatability. There are currently three recognised EFA, linoleic, α-linolenic and arachidonic acids all of which are polyunsaturated. Because of the complex nature of these compounds it is usual to designate their structure by the number of carbon atoms and double bonds they contain; thus linoleic acid which contains 18 carbon atoms and two double bonds is written 18:2. The EFA cannot be synthesised by the body and must therefore be supplied in the diet. Linoleic and α-linolenic acids are the parent compounds from which the more complex, longer chain compounds (derived EFA) can be made. The EFA are involved in many aspects of health including skin and coat condition, kidney function and reproduction. It is in the formation of the EFA that we see an important difference between the cat and other mammals, a contrast that is repeated for other nutrients. It has been reported that cats have only a limited ability to convert the parent EFA into the longer chain derivatives (Rivers, 1982); the lion appears to be similar in this respect. As a result cats require a pre-formed dietary source of 20:3 or 20:4 (arachidonic) acids which in practical terms means a requirement for EFA of *animal* origin.

In an elegant study of EFA in cats MacDonald *et al.* (1984a) concluded that dietary linoleic acid at 2.5 % of energy was probably adequate and that, given an ideal level of linoleate, the arachidonic acid requirement was not less than 0.04% of energy. However, the interrelationship between these two compounds means that a higher level of arachidonic acid in the diet will spare the need for linoleic acid. Conversely, the minimum requirement for arachidonic acid would be much more than 0.04% if linoleic acid was below the ideal level or absent from the diet altogether. In practical terms the EFA requirements of the cat are met by a combination of linoleic and arachidonic acids (the former being much more widely available than the latter) from a blend of vegetable and animal oils and fats in the food.

Protein and Amino Acids

All proteins are compounds of carbon, hydro-gen and oxygen but unlike carbohydrates and fats they always contain nitrogen. Most proteins also contain sulphur. Proteins are very large molecules which consist of chains of hundreds (or perhaps thousands) of much smaller sub-units called amino acids. Although there are only about 20 amino acids in proteins, the variety of sequences in which they can be arranged is almost infinite and this results in the wide variety of proteins which occur in nature. Animals need dietary protein to provide the specific amino acids that their tissues cannot synthesise at a rate sufficient for optimum performance. These amino acids are then reformed into new proteins which are essential constituents of all living cells where they regulate metabolic processes (in the form of enzymes) and provide structure and are therefore required for tissue growth and repair. Amino acids can be conveniently divided into two classes: essential (indispensable) and non-essential (dispensable). As their name suggests, the essential amino acids cannot be made by the body in sufficient amounts and must therefore be present in the food. The non-essential amino acids can be made from excesses of certain other dietary amino acids, although, as components of body proteins, they are as important as the essential varieties. Essential amino acid requirements for kitten and puppy growth and for adult dog maintenance are summarised in Appendix I.

Requirements for other species have yet to be determined. At WCPN we have conducted some preliminary investigations on amino acid requirements of the adult cat (Burger and Smith, 1990). Earlier work at WCPN has shown that when all essential amino acids are present at more than adequate concentrations, about 10% protein energy is required to maintain adult cats in protein (nitrogen) balance (Burger *et al.*, 1984). This value is higher than corresponding figures for other mammals including the dog and is another example of the nutritional speciality of cats. From these studies, it appears that the higher protein requirement of the cat is not due to an increased requirement for essential amino acids but a need for more total protein i.e. dispensable amino acids or protein nitrogen. This, in turn, appears to be due to the cat's inability to adjust amino acid breakdown even when receiving a low protein diet (Rogers and Morris, 1982a). The cat is also unusual in its dependence on the amino acid arginine. Arginine deficiency in the cat rapidly results in severe adverse effects because of an inability to metabolise nitrogen compounds (via the urea cycle), which then accumulate in the bloodstream as ammonia (hyperammonaemia) and in serious cases can lead to death within several hours. It seems that there is no other essential dietary component (including water) whose deficiency has such a drastic effect upon the animal. The rapidity of the effects is second only to a lack of oxygen. This unique requirement appears to be due to an inability to synthesise the amino acid ornithine (also a component of the urea cycle) since the latter protects cats against the adverse effects of arginine deficiency (Morris and Rogers, 1978). Although other animals require arginine for growth, in general they do not need it for adult maintenance. Those that do (like the dog) seem to be much less sensitive to a deficiency and have a much lower dietary requirement than cats.

Most if not all of the investigations into the protein and amino acid requirements of dogs and cats have been conducted using semi-purified or 'synthetic' diets, in which the protein level or amino acid profile has been adjusted for the purposes of the study. In extrapolating these results to practical feeding or design of diets it is important to allow for several factors. The essential amino acid profile of a given protein is of paramount importance. Few, if any, naturally occurring proteins would have the amino acid content of a specially made test diet. Furthermore, availability or digestibility of proteins will vary from one source to another and from one animal to another. Animal proteins generally have a more balanced amino acid profile and better digestibility than plant proteins. This whole subject represents a good example of the difference between a precise requirement, determined under carefully defined and controlled test conditions and a recommendation

which must apply to a very large number of animals eating a wide range of foods.

Despite these limitations, studies to determine the precise protein and amino acid requirements of the companion animals are important steps in refining the formulation of diets. Protein is a precious raw material and one which should be used as efficiently as possible.

Taurine

No discussion of the amino acid requirements of companion animals would be complete without at least a brief explanation of the importance of taurine. Strictly speaking, taurine is an amino sulphonic acid which is not part of the polypeptide chain of protein. It is an end-product of sulphur amino acid metabolism and is normally produced from the sulphur-containing amino acids, methionine and cystine. The particular importance of taurine in cat nutrition was first discovered less than 20 years ago when Hayes *et al.* (1975) showed that taurine was an essential nutrient for the cat and a deficiency was associated with central retinal degeneration. Unlike most other animals, cats cannot synthesise sufficient taurine to meet their needs and the special sensitivity of the cat is heightened by its total dependence on taurine for the formation of bile salts. Unlike other species, it does not also use glycine for this purpose (see Chapter 3). The cat needs, therefore, a dietary supply of taurine and the 'animal dependence' theme is again shown in this instance as taurine is found almost exclusively in animal-derived materials, little is present in plants.

Although the discovery of taurine function in the cat centred on retinal function, more recent research suggests that the importance of taurine in cat nutrition extends beyond this area. Sturman *et al.* (1986) reported that a taurine-free diet fed to queens during gestation and lactation resulted in poor reproductive performance typified by frequent foetal resorption, low birth weight of kittens, poor survival and a reduced growth rate. Abnormalities in neurological function and skeletal growth also occurred. The work of Pion *et al.*

(1987) suggests that taurine deficiency in cats is also linked to dilated cardiomyopathy—a degenerative disease of heart muscle. Yet another aspect of taurine has recently been the subject of extensive research. The cat's minimum requirement was originally investigated using semi-purified diets (Burger and Barnett, 1982; Rogers and Morris, 1982b). A dietary content of 500 mg/kg dry matter (approximately 24 mg per MJ) was identified as adequate for pregnancy in cats (NRC, 1986). However, Pion *et al.* (1989) subsequently reported that in commercial canned cat foods a much higher dietary taurine level was needed—2000 mg/kg dry matter—to maintain adequate plasma taurine levels. Earle and Smith (1991b) fed a specially-prepared canned food to cats and found that low plasma taurine occured if the diet contained less than this concentration. The authors suggested that reduced gastro-intestinal absorption and excessive excretion from the digestive tract were both possible reasons for the increased dietary requirement.

As was discussed in the previous section, great care is needed in transferring results from studies using semi-purified diets to commercial foods. Nevertheless pet food manufacturers ensure that their foods contain an appropriate amount of taurine supplementation and support this with feeding studies to confirm that plasma taurine values are satisfactory. Taurine has certainly been a major focus of attention during its relatively brief history in cat nutrition and continues to be the subject of much investigation.

Minerals

Calcium and Phosphorus

Calcium and phosphorus are closely interrelated nutritionally and will therefore be discussed together. They are the major minerals involved in the structural rigidity of bones and teeth. Calcium is also involved in blood clotting and in the transmission of nerve impulses. The level of calcium in the blood plasma is crucial to these functions and is very carefully regulated. Phosphorus also has many other functions (more than any other

mineral element) and a complete discussion of phosphorus metabolism would require coverage of nearly all the metabolic processes in the body. Phosphorus is involved in many enzyme systems and is also a component of the so-called 'high energy' organic phosphate compounds. These are mainly responsible for the storage and transfer of energy in the body.

The ratio of calcium to phosphorus in the diet is of great importance. The minimum calcium to phosphorus ratio for growth is generally considered to be about 1:1. For adult animals it is somewhat less critical. Imbalance in this ratio, where calcium is much less than phosphorus, leads to a marked deficiency of calcium in relation to bone formation. There is evidence that very high levels of these minerals or a very high ratio are also harmful. The metabolism of calcium and phosphorus is closely linked with vitamin D and this will be discussed later in the chapter.

Potassium

Potassium is found in high concentrations *within* cells and is required for nerve transmission, fluid balance and muscle metabolism. A deficiency causes muscular weakness, poor growth and lesions of the heart and kidney. Potassium is widely distributed in foods and naturally occurring deficiencies are rare, however its requirement is linked to protein intake so care may be needed in ensuring that high protein diets contain adequate potassium (see Chapter 4).

Sodium and Chloride

In contrast to potassium, sodium occurs mainly in the extracellular fluids, but like potassium, it is important for normal physiological function. With chloride, these substances represent the major electrolytes of the body water. Common salt (sodium chloride) is the most usual form of these two minerals added to food, so the dietary recommendation is often expressed in terms of sodium chloride. As with potassium, it is most unlikely that normal diets will be deficient in these two minerals.

Magnesium

Magnesium is found in the soft tissues of the body as well as in bone. Heart and skeletal muscle and nervous tissue depend on a proper balance between calcium and magnesium for normal function. Magnesium is also important in sodium and potassium metabolism and plays a key role in many essential enzyme reactions, particularly those concerned with energy metabolism. A deficiency of magnesium is characterised by muscular weakness and, in severe cases, convulsions. Nevertheless a dietary deficiency of magnesium is very unlikely. In contrast, very high intakes of magnesium in cats have been associated with an increased incidence of Feline Lower Urinary Tract Disease (Markwell and Gaskell, 1991).

Trace Elements

Iron

Iron is probably the best known trace element and much research has been carried out on its functions and requirements, particularly in the dog. Iron is a component of haemoglobin and myoglobin which play an essential role in oxygen transport; it is also an essential part of many enzymes (haem enzymes) which are involved in respiration at the cellular level, i.e. the oxidation of nutrients to form chemical energy. The absorption of iron is known to be influenced by a number of factors. Ferrous iron is better absorbed than ferric iron and iron contained in foods of animal origin tends to be better absorbed than that from vegetable sources. Some evidence from studies in man suggests that the inclusion of soy protein in a diet reduces the absorption of iron and other trace elements (zinc and manganese) and it may be important to ensure that the concentration of iron in products containing high levels of soy protein is always above the recommended allowance.

A deficiency of iron results in anaemia with the typical clinical picture of weakness and fatigue. Conversely iron, like most trace elements, is toxic if ingested in excessive amounts. Iron toxicity in dogs has been extensively studied and is associated with anorexia and weight loss. Of the iron salts investigated, ferrous sulphate was the most toxic, presumably because its absorption is high; iron oxide was much less toxic, because its bioavailability is low.

Copper

Copper is involved in a broad range of biological functions and is a constituent of many enzyme systems, including one which is necessary for the formation of the pigment melanin. Copper is very closely linked with iron metabolism and its deficiency impairs the absorption and transport of iron and decreases haemoglobin synthesis. Thus a lack of copper in the diet can cause anaemia even when the intake of iron is normal. Bone disorders can also occur as a result of copper deficiency and in this case the cause is thought to be a reduction in the activity of a copper-containing enzyme leading to diminished stability and strength of bone collagen.

Ironically *excess* dietary copper may also cause anaemia which is thought to result from competition between copper and iron for absorption sites in the intestine. Bedlington Terriers are known to display an unusual defect which results in toxic excesses of copper in the liver. The disorder results in hepatitis and cirrhosis and appears to be inherited. It has also been identified in other breeds including West Highland White Terriers and Dobermann Pinschers (Thornburg *et al.*, 1985a, b). For these particular breeds of dog it is probably a good idea to exclude foods with high copper contents and to avoid the use of copper-containing mineral supplements.

Manganese

Although little is known about the specific manganese requirements of companion animals, a considerable amount of evidence has accumulated that this trace element is essential in animal nutrition. Manganese is known to activate many metal–enzyme systems in the body and is therefore involved in a wide variety of reactions. A deficiency of manganese is characterised by defective growth and re-production and disturbances in lipid metabolism. These effects, like those of copper deficiency, are probably caused by inactivation or malfunction of one or more of the enzyme-catalysed reactions associated with these physiological processes. Although manganese is reported to be one of the least toxic of the trace elements, toxicity has been reported in several species, including cats, where it caused poor fertility and partial albinism in some Siamese. One of the other effects of excess manganese is on haemoglobin formation where its action is thought to be similar to that previously described for copper, i.e. competition with iron at absorption sites in the alimentary tract.

Zinc

The functions of zinc can be divided into two broad categories: enzyme function and protein synthesis. Zinc is required by all animals, but the zinc requirement is particularly affected by other components of the diet. For example, a high dietary calcium content or a vegetable protein-based diet can dramatically increase the zinc requirement and this latter effect may be related to that reported for iron absorption. Zinc availability is also decreased by the presence of phytic acid in the food. This compound is a complex organic molecule containing phosphorus which can bind trace elements such as zinc and thereby reduce their availability to the animal. Phytic acid and its derivatives (the phytates) are found particularly in cereals and related materials. In foods containing these, care must always be taken to ensure that the zinc concentration is adequate. For example, Van den Broek and Thoday (1986) reported signs of zinc deficiency in dogs receiving cereal-based dry diets which contained zinc at levels that were actually above the minimum requirement.

Zinc deficiency is characterised by poor growth, anorexia, testicular atrophy, emaciation and skin lesions. Although all nutrients are important, the link between zinc and skin and coat condition makes this trace element particularly crucial for the pet animal. This is because a marginal deficiency may occur where an animal is not obviously unwell but its skin or coat condition is sub-optimal and significantly detracts from its appearance. Zinc is relatively non-toxic. It interferes with the absorption and utilisation of iron and copper (especially the latter) so the severity of the effects of high intakes of zinc is dependent on the levels of these other trace elements in the diet. With normal dietary contents of iron and copper, it appears that zinc concentrations up to eight times the minimum requirement will not produce adverse effects.

Iodine

The only recognised function of iodine is in the synthesis of the thyroid hormones which are released by the thyroid gland and regulate the metabolic rate of the animal. One of the factors which influences the output of the thyroid hormones is the availability of sufficient iodine. In the absence of the requisite amount the thyroid gland increases its activity in an attempt to compensate for the iodine deficiency. As a result the gland (which is located in the neck region) enlarges and becomes turgid, a condition known as goitre which is the principal sign of iodine deficiency. Nevertheless there are other factors which are important in the occurrence of goitre. These include infectious agents, naturally occurring substances in the diet (goitrogens) which inhibit the synthesis, release or general effectiveness of the thyroid hormones, and genetically determined defects in the enzyme systems responsible for the biosynthesis of these hormones.

In man severe reduction in thyroid activity (hypothyroidism) is often referred to as cretinism when it occurs in children and myxoedema in adults. Hypothyroidism has been reported in dogs and iodine deficiency has also been observed in zoo felids, domestic cats, birds and horses. Clinical signs include skin and hair abnormalities, dullness, apathy and drowsiness. There can also be abnormal calcium metabolism and reproductive failure with foetal resorption. Excessive iodine intakes can be toxic. Hypothyroid cats given high doses of iodine (about 150 times the minimum requirement) were reported to show adverse effects which included anorexia, fever and weight loss (NRC, 1986). In other animals very large doses of iodine have been reported to produce acute effects similar to those of a *deficiency*. The high doses in some way impair thyroid hormone synthesis and can produce so-called iodine myxoedema or toxic goitre. Horses seem particularly sensitive to excessive iodine intakes, with maximum safe levels only one-tenth of those reported for other mammals.

Selenium

Ironically, attention was first focused on selenium because of its toxicity. The discovery that it is an essential nutrient for mammals took place fairly recently, about 35 years ago. Any discussion of the biochemical role of selenium has to take into account the close interrelationship of this element with vitamin E and the sulphur-containing amino acids methionine and cystine. The link with vitamin E is particularly important since one nutrient can 'spare' a deficiency of the other. Nevertheless it has been demonstrated in many animals that selenium cannot be replaced completely by vitamin E and has a discrete, unique function. Selenium is known to be an obligatory component of an enzyme called glutathione peroxidase which protects cells against damage by oxidising substances (notably lipid peroxides) which can be released by various metabolic processes in the body. Sulphur amino acids are required to form the enzyme: vitamin E is thought to act within the membranes, preventing oxidation of the lipids. In this way the functions of these three nutrients are closely linked.

The interactions of selenium are obviously highly complex and much is still unknown about this substance. It may, for example, be involved in processes unrelated to its role as a component of glutathione peroxidase. It has been reported to protect against lead, cadmium and mercury poisoning and has even been implicated as an anti-cancer agent in both experimental and epidemiological studies. Selenium deficiency has many effects but one described in dogs is degeneration of skeletal and cardiac muscles. Effects of deficiency in other species include reproductive disorders and oedema.

As mentioned earlier, selenium is highly toxic in large doses and the available evidence suggests that the difference between the recommended allowance and the toxic dose may be quite small. Injudicious supplementation of foods is therefore particularly dangerous in this respect.

Cobalt

Cobalt is a component of vitamin B_{12} and this may be its only biological function in the dog and cat. Under laboratory conditions cobalt can replace zinc in a few zinc-containing enzyme systems but whether this is of biological importance is not known. In the horse, vitamin B_{12} can be synthesised by caecal and colonic bacteria in the presence of cobalt. In the dog and cat, this synthesis may be of only limited importance. It is likely that to be of significant nutritional value, cobalt must be ingested by the dog and cat principally as vitamin B_{12}. With an adequate supply of the vitamin it is very doubtful whether any additional cobalt is required. Vitamin B_{12} will be discussed later in this chapter.

Other trace elements

A number of trace elements have been demonstrated to be necessary for normal health in mammals, although specific requirements have not been established for companion animals. These elements are listed in Table 2.1 with a brief summary of their functions. From work with other animals it appears that the amounts required in the diet are very low and the likelihood of a deficiency of any of these nutrients in a normal diet is consequently almost non-existent. Conversely, as with the majority of the trace elements these substances are all toxic if fed in large quantities, although the amounts which can be tolerated vary from one element to another. Arsenic, vanadium, fluorine and molybdenum are the

A SUMMARY OF THE FUNCTIONS OF SOME TRACE ELEMENTS	
Element	**Involvement**
Chromium	Carbohydrate metabolism, closely linked with insulin function
Fluoride	Teeth and bone development, possibly some involvement in reproduction
Nickel	Membrane function, possibly involved in metabolism of the nucleic acid RNA
Molybdenum	Constituent of several enzymes, one of which is involved in uric acid metabolism
Silicon	Skeletal development, growth and maintenance of connective tissue
Vanadium	Growth, reproduction, fat metabolism
Arsenic	Growth, also some effect on blood formation, possibly haemoglobin production

TABLE 2.1: A summary of the functions of some trace elements.

most toxic, whereas relatively large amounts of nickel and chromium can be ingested without adverse effects.

Vitamins

The vitamins may be conveniently divided into two sub-groups: fat-soluble and water-soluble. Apart from the obvious chemical difference the degree of storage in the body also differs, fat-soluble vitamins being stored to a greater extent than the water-soluble type. A regular daily supply is therefore less critical in the case of the fat-soluble vitamins.

Fat-Soluble Vitamins

Vitamin A. The term vitamin A is now used to describe several biologically-active compounds but retinol is the substance of primary importance in mammalian physiology. In nature vitamin A is found to a large extent in the form of its precursors, the carotenoids, which are the yellow and orange pigments of most fruits and vegetables. Of these, β-carotene is the most important 'provitamin A' because it has the highest activity on a quantitative basis, consisting essentially of two vitamin A-type molecules linked together, which most animals can convert to two molecules of the active vitamin.

Here we find yet another important difference between the cat and other mammals. It has been shown that the cat is unable to convert β-carotene to vitamin A; cats there-fore require a pre-formed dietary source of vitamin A of which the most common forms are derivatives of retinol (retinyl acetate and retinyl palmitate). The practical consequence of the peculiarity is that the cat must have at least some animal-derived material in its diet since *pre-formed* vitamin A compounds are not present in plants.

The best known function of vitamin A is in the physiological functions of vision. It is found in the retina combined with a specific protein called opsin. The compound is called rhodopsin (visual purple) and on exposure to light is split into opsin and a metabolite of retinol. It is the energy exchange in this process which produces nervous transmissions which are sent via the optic nerve to the brain and which result in visual sensations. Although the splitting of rhodopsin is reversible, a fresh supply of vitamin A is required to reform the visual pigment completely and so allow the process to continue. Vitamin A is involved in many other physiological functions one of the most important being the regulation of cell membranes; it it essential for the integrity of epithelial tissues and the normal growth of epithelial cells. Vitamin A is also involved in the growth of bones and teeth.

As might be expected, a deficiency of vitamin A has many far-reaching effects on the body and has been observed in many animals. The symptoms include xerophthalmia (excessive dryness of the eye), ataxia, conjunctivitis, opacity and ulceration of the cornea, skin lesions and disorders of the epithelial layers, e.g. the bronchial epithelium, respiratory tract, salivary glands and seminiferous tubules.

An *excess* of vitamin A is as harmful as a deficiency. A crippling bone disease with tenderness of the extremities associated with gingivitis and tooth loss has been described in cats given prolonged excessive doses of this vitamin either as vitamin A itself or by feeding large quantities of raw liver. Similar effects have been seen in dogs given large doses of vitamin A. Thus inclusion in the diet of foods containing large quantities of this vitamin, e.g. liver and the fish liver oils, must be very carefully controlled. Supplementation of an already adequate diet is not only unnecessary but potentially dangerous and should be avoided.

Vitamin D. There are several compounds which have vitamin D activity but the two most important are called ergocalciferol (vitamin D_2) and cholecalciferol (vitamin D_3). Both of these forms are effective as sources of vitamin D activity. There has been a large amount of research conducted on the metabolism of vitamin D in other mammals and it is now known that this vitamin undergoes a series of biochemical conversions in the kidney and liver before it becomes physiologically active. It is a dihydroxy derivative of the parent compound that is the most potent metabolite. Vitamin D is often called the 'bone vitamin' and its most clearly established function is to raise the plasma calcium and phosphorus levels to those required for the normal mineralisation of bone. In the small intestine vitamin D stimulates the absorption of calcium and phosphorus and is also involved in the mobilisation of calcium from bone to maintain a normal plasma calcium concentration. In fact the biochemical synthesis of the active vitamin D compound is triggered by a fall in plasma calcium. It is clear that the requirements for vitamin D are closely linked to the dietary concentrations of calcium and phosphorus and to the calcium/phosphorus ratio.

As vitamin D is involved in the absorption of calcium, it is most crucial during the growth and development of bone. i.e. in the young growing animal. A deficiency of this vitamin causes rickets. However there is evidence that most mammals can form vitamin D_3 from lipid compounds in the skin (provitamin D) in the presence of the ultra-violet component of sunlight and it is likely that adult animals need little if any *dietary* supply of this vitamin. Rivers *et al.* (1979) reported that cats are almost totally independent of a dietary source of vitamin D, even during growth and shielded from ultra-violet light, assuming they are fed a diet with adequate concentrations (and a correct ratio) of calcium and phosphorus. This appears to be because the cat mobilises stores of vitamin D_3 which it acquires during suckling. In contrast, Hazewinkel *et al.* (1990) reported that rickets in young dogs cannot be prevented or treated by ultra-violet radiation. This is due to a lack of provitamin D in the skin and means that dogs are dependent upon a dietary source of this vitamin.

As with vitamin A, excessive amounts of vitamin D cause adverse effects notably extensive calcification of the soft tissues, lungs, kidneys and stomach. Deformations of the teeth and jaws can also occur and death can result if the intake of the vitamin is particularly high. In fact, cholecalciferol is used as a rodenticide. Supplementation of vitamin D is therefore potentially hazardous and for cats, the requirement may be so low that any reasonable diet is bound to supply adequate quantities.

Vitamin E. The function which was first ascribed to this vitamin was that of preventing foetal resorption in animals that had been fed a diet containing rancid lard. The chemical name for this vitamin (tocopherol) is derived from the Greek word 'to bring forth offspring'. However in recent years studies on vitamin E have revealed much more of its role in the body although complete details of its function remain obscure.

It acts as an anti-oxidant and is important in maintaining the stability of cell membranes; in this its function is closely linked to that of the trace element selenium which was discussed earlier. The requirement for vitamin E also depends on the level of polyunsaturated fatty acids (PUFA) in the diet. Increasing PUFA increases the vitamin E requirement and this effect has been shown in many animals. It is difficult therefore to be precise about recommendations for dietary vitamin E; the levels stated in Appendix I are based on normal dietary concentrations of selenium and PUFA. Rancid fats should be avoided because they are particularly destructive to this vitamin.

Deficiency of vitamin E under experimental conditions presents a more bewildering range of physical abnormalities than is encountered with any other vitamin. These effects may be divided into four main areas: the muscle, reproductive, nervous and vascular systems. In dogs, a deficiency has been associated with one or more of these effects including skeletal muscle dystrophy, degeneration of the germinal epithelium of the testes and failure of gestation. Vitamin E deficiency in dogs has also been linked with impairment of the immune response. In cats inflammatory changes in body fat (steatitis–'yellow fat disease') occurs when low levels of vitamin E are fed in the presence of PUFA.

There is only very limited information on the effects of high vitamin E intakes. No deleterious effects were reported when about 10 times the recommended level was fed to weaned Beagle puppies for 15 weeks. However in other species some adverse effects on thyroid activity and blood clotting have been noted with high vitamin E intakes. The latter probably occurs via inhibition of vitamin K activity (see next section). Therefore high levels of this nutrient must be considered potentially harmful, although it is far from being as dangerous as vitamins A and D in excess.

Vitamin K. Vitamin K describes a group of compounds, the quinone derivatives, which regulate the formation of several factors involved in the blood clotting mechanism. A requirement for vitamin K has been demonstrated in the dog and it is unlikely that other animals are any different in this respect. Nevertheless the requirement in dogs was demonstrated under experimental conditions where the animals were made vitamin K deficient by the use of anti-coagulant drugs (such as the coumarin compounds) which antagonise the action of this nutrient. In normal healthy animals a vitamin K deficiency is very rare because most if not all of the daily requirement comes from bacterial synthesis in the intestine. It is only under abnormal conditions such as depression of bacterial synthesis (for example by drug treatment) and interference with the absorption or utilisation of vitamin K, that a dietary supply will be necessary. A diet containing only 60 $\mu g/kg$ dry matter (about 3.5 μg per MJ) was fed to adult male Beagles and cats for 40 weeks with no signs of deficiency although the same diet resulted in haemorrhages when fed to rats. A concentration of around 5 μg per MJ has been suggested as a minimum requirement for cats, although this is probably necessary only when bacterial synthesis has been suppressed or there are anti-vitamin K components in the diet. Very large intakes of vitamin K produce anaemia and other blood abnormalities in young animals but it does not appear to be particularly toxic.

Water-Soluble Vitamins

The water-soluble vitamins of major importance in companion animal nutrition are all members of the B complex and nearly all are involved with the utilisation of foods and the production or interconversion of energy in the body. In these processes, the B vitamins are used by the animal to form coenzymes (sometimes also called cofactors). These are relatively small organic molecules, associated with larger enzyme molecules, which are necessary for the enzymes to catalyse biochemical reactions effectively. The coenzymes often act by combining with and then releasing molecules or fragments of molecules, rather like a biochemical 'relay station'. Sometimes minerals and trace elements are also involved in these reactions as has been discussed earlier.

The B vitamins are now usually known by their chemical names, rather than by a letter/number combination, but this alternative nomenclature will be mentioned for vitamins where it is still in common use. In general some synthesis of most of the B vitamins occurs in the intestine of horses whereas dogs and cats usually need a dietary supply.

Thiamin (Aneurin, Vitamin B$_1$). Thiamin is a sulphur-containing compound which participates as a coenzyme in the form of its pyrophosphate (TPP), sometimes referred to as cocarboxylase. TPP is involved in several key conversions in carbohydrate metabolism and the thiamin requirement is dependent on the carbohydrate content of the diet. A high fat, low carbohydrate diet will spare the need for thiamin as less of this vitamin is required for fat metabolism than in carbohydrate utilisation.

Thiamin deficiency has been described in companion animals. Its primary effect is a 'biochemical lesion' resulting in impaired carbohydrate metabolism with abnormal accumulation of the intermediate compounds of the metabolic pathway. The deficiency expresses itself clinically as anorexia, neurological disorders (especially of the postural mechanisms) followed ultimately by weakness, heart failure and death. In man, thiamin deficiency is known as beri-beri. Thiamin is a particularly important vitamin from the aspect of dietary formulation because it is progressively destroyed by cooking and may also be inactivated by naturally occurring substances called thiaminases which are found in a number of foods, particularly raw fish and certain plants. Thiaminases are themselves inactivated by heat so the maintenance of an adequate thiamin intake must take all of these various factors into consideration. For commercially prepared heat-processed foods, the normal practice is to supplement with a large enough quantity before processing so that even if particularly serious losses occur, the amount remaining in the finished product will still meet or exceed the dietary recommendations.

Like the other water-soluble vitamins, thiamin is of low toxicity. Although intravenous *injection* of thiamin in dogs produces death through depression of the respiratory centre, the *oral* intake needed to cause the same effect is some 40 times the intravenous dose and represents a level many thousands of times the recommended dietary concentration.

Riboflavin (Vitamin B$_2$). Riboflavin is a yellow crystalline compound which shows a characteristic yellow-green fluorescence when dissolved in water. Riboflavin is a constituent of two coenzymes, riboflavin 5-phosphate and a more complex chemical called flavin adenine dinucleotide. These coenzymes are essential in a number of oxidative enzyme systems. Cellular growth cannot occur in the absence of riboflavin.

Riboflavin deficiency is associated with eye lesions, skin disorders and testicular hypoplasia. There is some evidence that part of the requirement for riboflavin can be met by bacterial synthesis in the intestine and that this is favoured by a high-carbohydrate, low-fat diet. However, in dogs and cats the daily needs for the vitamin are certainly greater than any possible contribution by this route so a regular dietary intake is necessary.

Pantothenic Acid. This substance is a constituent of coenzyme A which is an essential component of enzyme reactions in carbohydrate, fat and amino acid metabolism. A need for pantothenic acid has been demonstrated in dogs and cats. There are many deficiency signs, including depression or failure of growth, development of fatty liver and gastrointestinal disturbances including ulcers. In dogs, but not cats, alopecia has also been observed. These deficiency signs were produced using semi-purified diets. Under normal circumstances, using a mixture of foods, a deficiency of pantothenic acid is extremely unlikely; it is very widespread in animal and plant tissues, as implied by its name which means 'derived from everywhere'.

Niacin. Niacin is a generic name for two compounds with equal vitamin activity, nicotinamide and nicotinic acid. It is a component of two very important coenzymes, the nicotinamide adenine dinucleotides, which are required for oxidation-reduction reactions necessary for the utilisation of all the major nutrients. In mammalian species the requirement for niacin is influenced by the dietary level of the amino acid tryptophan, which can be converted to the vitamin. In cats this conversion does not occur but, unlike other differences in the cat, this is not due to lack of an enzyme. It occurs because the reaction sequence for the breakdown of tryptophan can go one of two ways and in the cat the enzyme responsible for the alternative 'non-niacin' pathway has a very high activity and effectively abstracts the tryptophan metabolites from niacin synthesis. This alternative pathway eventually breaks down the metabolites to supply energy, similar to the utilisation of carbohydrates.

Niacin deficiency has been described in dogs and cats and is accompanied by inflammation and ulceration of the oral cavity with thick, blood-stained saliva drooling from the mouth and foul breath. The deficiency syndrome is referred to as blacktongue in the dog and pellagra in man. Niacin is sometimes called the pellagra-preventing vitamin or PP factor. Large doses of nicotinic acid (but not nicotinamide) produce a flushing reaction in many animals including dogs. Thus if large therapeutic quantities of this vitamin need to be administered, it is preferable to use the amide form. Nevertheless, neither of the two forms of this vitamin could be described as highly toxic.

Pyridoxine (*Vitamin B₆*). There exist three related compounds under this heading with essentially equally effective activity: pyridoxine, pyridoxal and pyridoxamine. All three occur naturally and are interconvertible during normal metabolic processes. The biologically active compound is pyridoxal and the coenzyme form is pyridoxal 5-phosphate, which is involved in a large number and variety of enzyme systems almost entirely associated with nitrogen and amino acid metabolism. In fact pyridoxal is considered essential for practically all enzymic interconversions and non-oxidative degradations of amino acids. Some of these reactions have already been discussed in relation to other nutrients; for example, the synthesis of niacin from tryptophan involves this vitamin. As might be expected, a high protein diet exacerbates vitamin B_6 deficiency, an effect which is comparable with the effect of high carbohydrate diets on thiamin deficiency.

A deficiency of pyridoxine results in weight loss and a type of anaemia. In cats, irreversible kidney damage can also occur with tubular deposits of calcium oxalate crystals (pyridoxine is required in the conversion of oxalate to glycine). Dermatitis and alopecia have occasionally been reported in pyridoxine deficiency in dogs. Like the other water-soluble vitamins, pyridoxine and its derivatives are not considered highly toxic.

Biotin. Like other B vitamins, biotin is presumed to function as a coenzyme and is necessary for certain reactions involving the metabolism of the carboxyl (CO_2) group which is initially bound to biotin before transfer to an 'acceptor' molecule. In biotin deficiency there is a reduction in amino acid incorporation into proteins, apparently due to a fall in the synthesis of dicarboxylic acids. Impairment of glucose utilisation and fatty acid synthesis have also been reported. In the early stages of deficiency the principal clinical sign seems to be a scaly dermatitis. Although these effects were initially investigated using other animals, it is now known that biotin is required by dogs and cats and similar deficiency signs have been described. However, it is very difficult to produce biotin deficiency with a normal diet because most, if not all of the daily requirement can be met from synthesis by the gut bacteria. Deficiency signs have been produced in the dog and cat only when antibiotics were given to suppress bacterial action and large quantities of whole egg white were included in the diet. Egg white contains a protein called avidin which forms a stable and biologically inactive complex with biotin. Avidin will also 'neutralise' biotin produced by bacteria. Avidin itself is relatively heat sensitive, so if eggs constitute a significant proportion of the diet they should be fed cooked rather than raw. It is also important to realise that antibiotic drugs can increase the requirement for vitamins like biotin (see also vitamin K and folic acid) because they destroy the intestinal bacteria responsible for their manufacture. Nevertheless the likelihood of a naturally occurring biotin deficiency is remote.

Folic Acid (*Pteroylglutamic Acid, Folacin*). Folic acid is usually found in nature in the form of conjugates with the amino acid, glutamic acid. The biologically active coenzyme is the tetrahydro derivative, often abbreviated as THFA or FH_4. There are several other forms of THFA with coenzyme activity all of which are usually grouped under the generic name of the folates or folate coenzymes. The folates are involved in the transfer of single carbon groups (e.g. methyl and formyl) which are important in several ways. Perhaps the most significant reactions are those necessary for the synthesis of thymidine, an essential component of the nucleic acid DNA. Lack of an adequate supply of DNA prevents normal maturation of primordial red blood cells in bone marrow and the typical signs of folic acid deficiency are therefore anaemia and leukopenia. Folic acid deficiency has been described in dogs and cats but usually only when semi-purified diets were fed in the presence of antibiotics. It is likely that most of the daily requirement for folate is met by bacterial synthesis in the intestine.

Vitamin B₁₂. This vitamin is unique in being the first cobalt-containing substance shown to be essential for life and is the only vitamin that contains a trace element. Vitamin B_{12} is also known as cobalamin but is usually isolated in combination with a cyanide group linked to the cobalt atom. This form is known as cyanocobalamin and is sometimes used as a synonym for vitamin B_{12} itself. The active coenzyme is yet another derivative where a new chemical group replaces cyanide in the parent molecule. Like the folates, vitamin B_{12} is involved in the transfer of single carbon fragments and its function is closely linked to that of folic acid itself.

Vitamin B_{12} is also involved in fat and carbohydrate metabolism and in the synthesis of myelin, a constituent of nerve tissue. The typical signs of a B_{12} deficiency in many ways resemble those of folate deficiency but characteristically also involve neurological impairment as a result of inadequate production of myelin. Vitamin B_{12} is only poorly absorbed from ingested food unless a protein, 'intrinsic factor', is present in the intestine. This factor presumably facilitates transfer of the vitamin across the mucosal membrane. Failure to absorb B_{12} due to lack of intrinsic factor, results in pernicious anaemia with neurological degeneration.

These effects have been described in other animals including man, but less information is available for the dog and cat. The vitamin has been shown to be needed by these two species but a quantitative requirement has not been determined in detail. The stated requirements are based on some research conducted in dogs and cats and on data from other mammals.

Choline. Choline is a component of the phospholipids which are essential components of cell membranes; it is the precursor of acetylcholine, one of the body's neurotransmitter chemicals. It is an important methyl donor, that is it supplies single carbon fragments for metabolic conversions, the significance of which have already been discussed in the previous sections dealing with folic acid and vitamin B_{12}. A deficiency of choline causes several abnormalities including kidney and liver dysfunction which in the dog and cat are usually manifested as fatty infiltration of the liver. The precise mechanism for this is not known but may be linked to inadequate biosynthesis of specific types of phospholipids, leading to impaired rates of lipid transport.

The requirement for choline in the diet can be modified by a number of factors, in particular the dietary concentration of methionine. Since methionine can also act as a methyl donor in intermediary metabolism, an increased dietary supply of one tends to spare the need for the other. Some work with cats has shown that methionine can completely replace the dietary need for choline if supplied in adequate amounts (Anderson *et al.*, 1979). In view of the sparing effect of methionine and the widespread distribution of choline in plant and animal materials it is most unlikely that a dog or cat will become choline deficient under normal circumstances.

Ascorbic Acid (*Vitamin C*). Most mammals do not need a dietary source of this vitamin because they are able to synthesise it from glucose. The main exceptions are man and the other primates and the guinea pig. Most birds can synthesise vitamin C but fish require a dietary source. Some researchers have claimed that a number of diseases in the dog can be ameliorated by ascorbic acid. Furthermore, skeletal diseases such as hypertrophic osteodystrophy, hip dysplasia and a number of others, particularly those common in the large and giant breeds, have been said by some to resemble ascorbic acid deficiency (scurvy). However, other groups have consistently failed to show any benefits of vitamin C in either alleviating or preventing these diseases. There are reports that extra vitamin C may be beneficial for dogs and horses which are highly active or under stress, e.g. under a hard training regime or in harsh conditions (such as sledge dogs). It is also possible that some individuals may have a reduced capacity to synthesise this vitamin but on the available evidence it seems that there is no *general* dietary requirement for ascorbic acid in normal healthy companion animals, other than fish.

Dietary Recommendations

In this chapter much emphasis has been given to the fact that nutrient requirements are bounded by a minimum and, in some cases, a maximum value. In other words the amount of nutrient needed in a diet must lie on a 'plateau' between deficiency on one hand and toxicity on the other. The main criteria for what constitutes a complete diet can therefore be summarised thus:

- the content of each nutrient must be on the plateau
- each nutrient must be present in the correct ratio to the energy content of the diet
- each nutrient must be at the correct ratio to other nutrients (where appropriate)
- each nutrient must be in a form that is usable by the animal for which the diet is made.

At the start of this chapter we compared minimum requirements (MDR) with recommended allowances (RDA). Compilation of values for the nutrient content of diets can be made using either of these concepts. Appendix I contains a number of lists of nutrient contents and some, like those of the National Research Council, are based on studies with semi-purified diets which mean that they represent MDR values. Other nutrient profiles, including those developed by nutritionists at WCPN, incorporate some allowances for availability and animal variability and therefore represent a form of RDA expressed as a dietary concentration. These figures are *not* a guarantee of nutritional quality but are good guidelines to produce a correctly balanced 'practical' diet. As was stated earlier, MDR data should be used to derive RDA which can then be applied to all animals within a given population. Nevertheless the final arbiter of nutritional adequacy is the animal itself and the ultimate endorsement of these dietary values can only come from feeding studies.

The general nature of this chapter should not obscure a particularly fascinating facet of companion animal nutrition, namely the atypical metabolism of the cat. This pet animal depends on a supply of at least some animal-derived materials in its diet and must be regarded as an obligate carnivore. Why should it not possess fully active enzyme systems responsible for the production of taurine, EFA and vitamin A and the conservation of protein? Is it possible that during the course of evolution these functions have been lost because of the ability of the cat family to catch prey and live on what is an almost entirely animal-based diet? Alternatively, were the early mammals obligate carnivores and the cat family represents an early branch of the evolutionary tree with such an efficient predatory lifestyle that it was subjected to little or no environmental pressure to utilise plant materials? Perhaps an investigation of the nutritional requirements of a primitive placental mammal like the hedgehog would reveal some interesting facts to contribute to the discussion!

Whatever the reasons for these differences between the dog and cat it must always be remembered by those involved in any aspect of pet feeding that, nutritionally and biochemically, a cat is not just a small, highly agile dog that climbs trees.

CHAPTER 3

Digestion and Absorption

IAN E. MASKELL and JANEL V. JOHNSON

Introduction

It is clear from the previous chapter that foods comprise a diversity of organic constituents, many in the form of large insoluble molecules. These must be degraded to simple molecular compounds before they can traverse the intestinal mucosa and enter the general circulation for delivery to specific sites within the body. The process of degradation is termed 'digestion' and passage across the intestinal mucosa 'absorption'. The coalition of these processes is central to the theme of nutrition; a diet with an ideal nutrient profile and optimal palatability is of no nutritional benefit if it cannot be broken down and assimilated following consumption.

Digestion involves a combination of mechanical, chemical and microbial activities all contributing to the sequential degradation of food components. Mastication and alimentary muscular contractions mechanically diminish the size of ingested food particles. Enzyme rich digestive juices secreted into digesta in the stomach and small intestine instigate chemical degradation. Bacteria inhabiting the terminal section of the alimentary canal also produce enzymes capable of chemical digestion. The significance of the so-called 'hindgut microflora' depends on both the specific characteristics of the diet and the species concerned. The horse for example is capable of considerable 'hindgut bacterial fermentation', and indeed relies heavily on the phenomenon for digestion of its normal vegetative diet. The cat, conversely, relies little on bacterial activity to sustain its carnivorous lifestyle.

The scope of this text is broad and the species concerned exhibit considerable diversity. The object of this chapter is **not** to delineate the intricacies of each species but rather to introduce general principles of digestion and absorption. The dog shares many similarities in its digestive physiology with humans, pigs and rodents—all widely studied monogastric omnivores. In general this chapter considers the dog as a yardstick, highlighting the important areas of difference in the other species. The cat is a carnivore which explains many of the fundamental differences when compared with dogs. Horses are monogastric herbivores, birds are diverse but those of interest in this book can be considered as monogastric 'grainivores' (i.e. seed eating) and fish are monogastric carnivores, omnivores and herbivores. The significance of both anatomical and biochemical differences between the species are emphasised where appropriate.

The chapter takes a systematic approach to

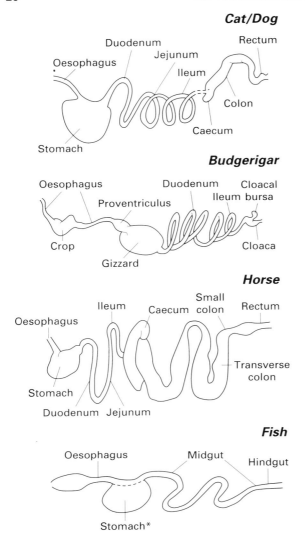

Cat/Dog

Duodenum
Rectum
Jejunum
Oesophagus
Ileum
Colon
Caecum
Stomach

Budgerigar

Oesophagus
Duodenum
Cloacal
Ileum bursa
Proventriculus
Crop
Cloaca
Gizzard

Horse

Small
Ileum
Caecum
colon
Rectum
Oesophagus
Transverse
colon
Stomach
Duodenum Jejunum

Fish

Oesophagus
Midgut
Hindgut
Stomach*

FIG 3.1: A diagrammatic representation of the digestive tracts of the cat/dog, budgerigar, horse and fish.* See page 87.

digestion, tracing the route of food ingested by the mouth, travelling via the oesophagus to the stomach, through the small and large intestine and finally, excretion of the undigested residue. The key processes of digestion and absorption are described at each stage. The final section deals with water balance.

Control of Digestion

The actions of the gastrointestinal tract are under both voluntary and involuntary control. Ingestion, chewing and swallowing are events consciously controlled by the individual, thereafter the digestive functions that commence in the back of the mouth with the initiation of swallowing are all reflex events, i.e. not under voluntary control, the only other *voluntary* event being control over the anal sphincter. As food passes from the pharynx via the oesophagus to the stomach, sphincters open and close under involuntary nervous control emanating subconsciously from the brain. As food enters the stomach, the reflex response of the gastric muscles is to relax in order to counteract undue increases of intragastric pressure. All digestive secretions in the stomach and intestines are controlled by nervous and hormonal interactions as is the motility of the tract which propels digesta in peristaltic waves towards the terminal end of the gut.

Comparative Digestive Anatomy

The digestive tract can be considered as a continuous tube differentiated into regions by structure and function. In mammals, birds and fish functionality of the tract is similar by virtue of the fact that all diets comprise a core of major nutrients common to all foods and hence the processes of digestion and absorption are analogous. Anatomically, however, the generas differ greatly, with mammals being differentiated further according to their digestive anatomy into ruminants (e.g. cattle and sheep) and simple stomached animals (e.g. dogs, cats and humans). Figure 3.1 illustrates anatomical differences of the dog/cat, bird (budgerigar), horse and fish digestive tracts. Macroscopically the tracts of dogs and cats are similar, the cat having a slightly less developed caecum. The crop, proventriculus (glandular stomach) and gizzard (muscular stomach) are unique to avian anatomy. The budgerigar is amongst those avians with no caeca, this feature is species specific and may be a weight reducing adaptation more important in flying birds (see Chapter 6). The equine tract has a comparatively small stomach and

INTESTINE LENGTHS AND AVERAGE DIGESTA TRANSIT TIMES				
	Dog	**Cat**	**Man**	**Horse**
Small intestine (m)	3.9	1.7	7.0	20
Large intestine (m)	0.6	0.4	1.8	7
Total length (m)	4.5	2.1	8.8	27
Body length (m)	0.75	0.5	1.75	3
Total length: body length ratio	4–5	3	5	9
Mean retention time (hr)	22.6 ± 2.2	13	45.6 ± 11.1	37.9 ± 5.3

TABLE 3.1: Intestine lengths relative to body length and average digesta transit times; typical values for dog, cat, horse and man.

well developed caecum and colon, these adaptations are consistent with horses being hindgut fermenting herbivores (see Chapter 8). Table 3.1 compares the length of the small and large intestines of dogs, cats, humans and horses with overall body length. Length is important as it influences the amount of time food actually resides within the gut and consequently the duration of digestion. The ratio of intestinal length to body length is greater in herbivores than carnivores, since vegetative foods require prolonged digestion; in omnivores the ratio is intermediate. Table 3.1 also shows average digesta transit times (Warner, 1981), these figures provide a guide but are subject to considerable variation. Diet, frequency of feeding, ambient temperature, pregnancy, exercise and age can all affect transit time.

The mammalian gut has a large absorptive surface area, which appears to be the main digestive adaptation enabling mammals to sustain faster digestive rates than more primitive animal groups like reptiles, for example. Increased absorptive efficiency is achieved mainly by increasing the intestinal surface area by a factor of about seven through folding to produce small finger like projections called microvilli (see Fig. 3.2). High surface area to bodyweight ratios in small mammals have been associated with a carnivorous diet, and lower values with a herbivorous diet (Barry, 1976) . The dog and cat have a similar surface area per cm of intestine length (jejunum, 54 vs 50 cm² and ileum, 38 vs 36 cm², respectively). However, although the overall intestine length in relation to body length is the lowest of the mammalian species discussed here, the cat has a greater potential absorptive capacity than the dog (Wood, 1944).

The gastrointestinal tracts of the dog and cat, whilst basically similar, do exhibit some characteristic differences. The gastric mucosa of the feline stomach is uniform in comparison with the canine stomach which has two distinct areas: the proximal stomach has a thinner mucous membrane with distinct gastric glands whilst the distal stomach has a thicker mucous membrane and less distinct glands (Nickel, Schummer and Seiferle, 1979). The main anatomical difference between the dog and cat is the caecum; in the dog the caecum exists as a diverticulum (blind ended sac) of the proximal colon and is more developed than in the cat, which has little more than a vestigial appendix.

The Mouth and Oral Digestion

Processing food before ingestion may have chemical and physical impacts by, for example, softening meat fibres, hydrating polysaccharides or gelatinising starch, but true

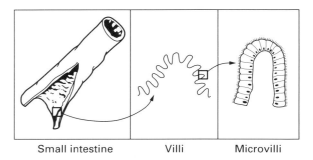

Small intestine   Villi   Microvilli

FIG 3.2: The method by which a large surface area for absorption is provided.

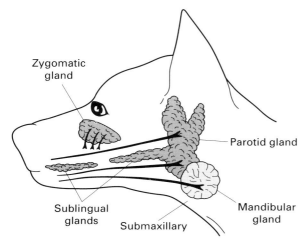

Zygomatic
gland

Parotid gland

Sublingual
glands

Submaxillary

Mandibular
gland

FIG 3.3: A diagrammatic representation of the anatomical
location of salivary glands and ducts in the dog.

digestion only commences when food enters the digestive tract.

Digestion in the mouth is mainly mechanical, mastication breaks down large fragments and mixes food with saliva. Saliva is secreted into the mouth by four pairs of salivary glands; the parotids in front of each ear, the mandibular (or submaxillary) glands on each side of the lower jaw, the sublingual glands under the tongue and the zygomatic gland located in the upper jaw below the eye. Figure 3.3 shows the anatomical location of salivary glands and related ducts in the dog, the glands are located similarly in the cat and horse. Saliva is usually present in the mouth but its flow is increased by the sight and smell of food. This effect, known as a gustatory response, was first examined in the classic 'bell' experiments for which Pavlov received a Nobel prize in 1904. Salivation continues when food enters the mouth and the effect is reinforced by chewing. Saliva is about 99% water, the remaining 1% comprising mucus, inorganic salts and enzymes. Mucus acts as an efficient lubricant and aids swallowing, particularly of dry foods. Unlike humans, the dog, cat and horse appear to lack the starch digesting enzyme α-amylase in their saliva, precluding an early commencement of starch digestion in the mouth, oesophagus and for a short period in the stomach. The lack of this enzyme is concomitant with observed behav-

iour of dogs which tend to bolt all but the toughest foods and the obligate carnivore nature of cats which tend to consume a low starch diet. In addition it is speculated that by dissolving some food particles and bringing them into intimate contact with taste buds, saliva plays a role in the perception of flavour and hence palatability.

Dentition closely reflects the normal diet of different species and indicates their natural eating behaviour and preferred diet. True carnivores do not have grinding molars and strict herbivores have no piercing canines. Omnivores have a diverse array of teeth intended to facilitate chewing, grinding and piercing— a testimony to their naturally diverse diet. The comparative dentition of the dog, cat and horse is outlined in Table 3.2 and illustrated in Fig. 3.4. Cats and dogs have the same number of incisor and canine teeth (six incisors and two canines in each jaw); dogs, however have 42 permanent teeth compared with 30 in the adult cat. Dogs have four premolar teeth on each side of the upper and lower jaw, and two molars on the upper and three molars on the lower jaw. Cats only have three premolars and one molar on the upper jaw and two premolars and a molar on the lower jaw, on each side. Both cats and dogs have an enlarged upper 4th premolar or carnassiate tooth and lower 1st molar, the combination of these 'carnassials' generates a powerful scissor action, highly effective for cutting raw flesh. Distinctive from cats, dogs have crushing molars indicating a capacity to utilise plant material; hence the dental formula of dogs suggests an omnivorous diet whilst that of the cat is consistent with a strictly carnivorous diet (as discussed in Chapter 2). The equine dental formula is not dissimilar to that of the dog, the horse has fewer premolars and more molars. Anatomically, however, the layout, shape and relative size of the teeth are very different (Fig. 3.4). The horse has large incisors designed specifically for nibbling and cutting grass; the canines (called tushes) are rudimentary in males and absent in females. There is a considerable gap between the tushe and the cluster of premolars and molars, the large grinding teeth located in the back of the

DOG

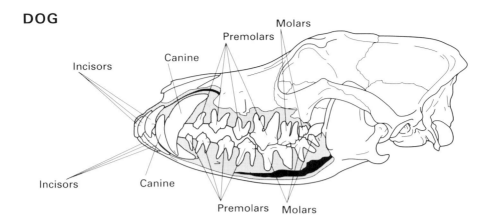

CAT

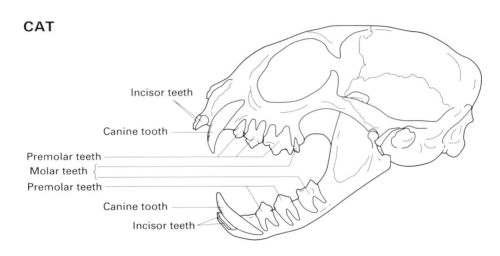

HORSE

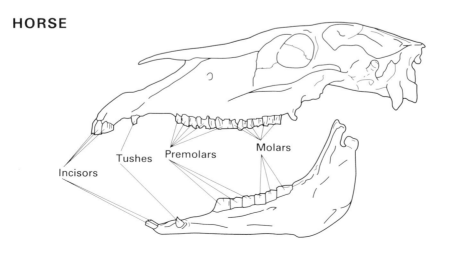

Fɪɢ 3.4: Pictorial anatomy of the jaw (not to scale).

COMPARATIVE DENTITION OF DOG, CAT AND HORSE					
	Incisors	Canines	Premolars	Molars	Total
Dog	12	4	16	10	42
Cat	12	4	10	4	30
Horse	12	4*	12/14	12	40/42

TABLE 3.2: Comparative dentition of the dog, cat and horse. * Absent in females.

mouth. Because a regular chewing action is inadequate to grind tough, course foods the horse has developed a lateral grinding movement which causes the flat surfaces of the upper and lower teeth to abrade with a strong, destructive motion.

Oesophagus

Food formed into a bolus by the tongue is transferred by swallowing to the oesophagus, a comparatively short, muscular tube leading from the mouth to the stomach. No digestive enzymes are secreted here but oesophageal cells do produce mucus to lubricate the process of peristalsis, automatic wavelike contractions and relaxations which propel food along the digestive tract which are stimulated by the presence of food. This process takes only a few seconds to transport food from the mouth to the stomach. The oesophagus enters the stomach via a ring of specialised muscle cells known as the cardiac sphincter. Normally in the contracted state, the cardiac sphincter is stimulated to relax by the approach of a peristaltic wave, allowing food to pass into the stomach. Pressure from the stomach, however, does not stimulate the sphincter to relax, thereby preventing reflux under normal circumstances. Vomiting is a specific reflex, during which the cardiac sphincter allows food to be expelled from the stomach under specific control of the vomiting centre in the brain. Dogs have a particularly well developed vomiting centre which may be stimulated by environmental changes as well as gastric and other clinical conditions. Consequently, vomiting is quite common in dogs, acting as an ultimate defence mechanism by which potentially toxic materials can be easily expelled from the gut after ingestion. This is a particularly useful routine in individuals with a tendency to scavenge.

Stomach

The stomach acts as a receptacle located at the beginning of the gastrointestinal tract. It has several functions. It stores food temporarily and controls the rate of entry of ingesta into the small intestine. The stomach participates in the initial stages of digestion by secreting acid and a precursor to the important proteolytic enzyme, pepsin. Gastric muscles control motility to ensure movement of food in an aboral direction (away from the mouth) and aid digestion by promoting mixing and grinding.

The mammalian stomach can be divided into five regions, cardia, fundus, body, antrum and pylorus (Fig. 3.5). The cardia is the point of entry of the oesophagus. The fundus forms a large blind pocket behind the cardia. The body forms the large middle section above the antrum which forms the distal region of the stomach with the pylorus at the terminal end, where the pyloric sphincter opens into the duodenum. The boundaries between these regions are arbitrary and functionally the stomach can be viewed as consisting of a proximal (corpus) and distal (antral) section. The proximal stomach is capable of expanding during the temporary storage of food, allowing consumption of discrete meals rather than a continual intake of small amounts. This is more significant in dogs which tend to be meal feeders rather than cats which tend to 'snack' feed. Gastric juices are also secreted in the proximal region. The distal region regulates release of hydrochloric acid, which sustains the environment at a low pH necessary for optimal enzyme activity.

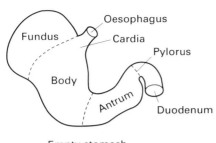

Empty stomach

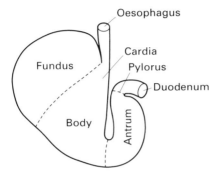

Full stomach

FIG 3.5: Diagram of the cross-section of the stomach showing anatomical and functional regions. Top: Empty stomach. Bottom: Full stomach, showing that increase in size is due to changes in the proximal part of the stomach.

The muscular structure of the stomach enables mechanical grinding of food particles and controls gastric motility. The pyloric sphincter both regulates the transit of solid food into the duodenum and prevents reflux.

Secretions in the Stomach

Gastric secretion is controlled by complex neural and hormonal interactions that dictate secretion at controlled rates and at the proper time. Events occurring in several parts of the gastrointestinal tract as well as the stomach and other systems of the body all play important roles in regulation of normal gastric secretion. Control is highly organised so that secretions between the main digestive phases are minimal whilst secretion during digestion is specific to the quantity and type of nutrients present. The mechanisms controlling acid secretion occur at three sequential times during digestion.

(i) *Cephalic phase*. The anticipation, sight, taste, smell and chewing of food stimulate the initial reflex response mediated in the brain. This neural mechanism primarily stimulates secretion of pepsinogen although small amounts of gastrin and hydrochloric acid are also released.

(ii) *Gastric phase*. The most important determinants of gastric secretion are events occurring in the stomach. Gastrin is the principal secretion during the gastric phase, its release stimulated by the presence of certain nutrients, specifically amino acids and mechanical distension of the gastric mucosa as food arrives via the oesophagus. Gastrin stimulates hydrochloric acid, pepsinogen and mucus secretion. Gastrin release is strongly inhibited when the pH in the antrum drops below 3.0; this is an important mechanism in the control of hydrochloric acid secretion. Gastrin secretion may also be controlled by peptide hormones (e.g. secretin or enteroglucagon) released during the intestinal phase of digestion and absorption may also produce some inhibition. It is speculated that the fundus may play a role in the inhibition of gastrin secretion although a mechanism has yet to be elucidated.

(iii) *Intestinal phase*. Food in the small intestine stimulates hydrochloric acid secretion in the stomach, mediated by both mechanical distension of the intestine and chemical stimulation by amino acids and peptides. In general stimulation mediated lower down the gastrointestinal tract tends to be inhibitory. For example the presence of fat in the small intestine stimulates release of the duodenal hormone enterogastrone, which inhibits gastrin release in the stomach.

Digestion in the Stomach

Food entering the stomach is mixed with gastric juice and subjected to further mechanical breakdown by stomach contractions. Gastric juice is produced by the gastric mucosa in response to the same stimuli that increase saliva flow. Gastric secretions play a role in the initiation of protein digestion, intestinal absorption of calcium, iron and vitamin B_{12} and maintenance of normal bacterial flora in the gastrointestinal tract as a whole through stabilisation of the luminal environment. The major component of gastric secretions is pepsinogen, an inactive precursor of the proteolytic enzyme pepsin, which allows the enzyme to be stored without damaging surrounding tissue. Hydrochloric acid converts the precursor to its active form by cleavage of small peptides from the pepsinogen molecule. Once formed, pepsin also converts pepsinogen to pepsin, the process is thereby termed autocatalytic. Pepsin acts on ingested food proteins by splitting them at specific points in their structure, the products formed are polypeptides of intermediate size, some of which may be potent simulators of gastric secretion (see below); subsequent protein digestion is initiated by pancreatic enzymes secreted into the duodenum.

Pepsin exhibits optimum activity at pH 2.0 maintained by gastric secretion of hydrochloric acid; its proteolytic activity ceases when chyme (partially digested food) leaves the stomach, since it is irreversibly inactivated at neutral pH. Pepsinogen secretion is not essential for digestion and absorption of a meal. Pancreatic proteases are important for completion of protein degradation and can cope adequately in the absence of pepsin if necessary. Since pepsin is most active when digesting collagen, its activity is more important for initiating the digestion of meat rather than vegetable protein. As such, pepsin is probably more important to the cat than the dog. There is speculation that the major role of pepsin is to yield 'bioactive' peptides that are more potent than their dietary precursors as stimulants of gastrin.

Although a range of other enzymes are found in gastric secretions none are thought to be significant in normal gastric activities.

The stomach also secretes a mucoprotein called intrinsic factor required to bind vitamin B_{12} so that absorption can occur in the small intestine. Intrinsic factor has been identified in gastric secretions of cats and dogs (Batt and Horadagoda, 1986). Malabsorption of B_{12} in humans is characterised by anaemia, and may be indicative of an atrophic gastritis, however neither are recognised as problems in the dog or cat. The rate of secretion of intrinsic factor increases with that of hydrochloric acid, however, sufficient is secreted under resting conditions to bind all B_{12} to be absorbed.

Gastric juice contains viscous mucus comprised mainly of glycoproteins and mucopolysaccharides. Mucus is a gel and forms a continuous protective covering of the epithelial cells of the stomach; its primary function is to stabilise the microenvironment of the mucosal surface. It also protects the gastric mucosa against the effects of acid and pepsin. Although not digested by pepsin, mucus is produced continuously in order to maintain its protective function. Mucus has been shown to accelerate healing of gastric ulcers, and prevent recurrence of duodenal ulcers. Mucus gel lining the mucosal surface acts as a fine meshwork that retards the passage of macromolecules but permits the movement of electrolytes and other small molecules. This selective permeabilty also helps to provide a barrier against micro-organisms.

Small Intestine and Pancreas

The small intestine is so called because of its narrow bore not its length; although its diameter is much less than that of the large intestine it is in fact several times longer. Enzymic digestion is completed in the small intestine, i.e. all digestible protein, fat and carbohydrate is reduced to amino acids, dipeptides, glycerol, fatty acids and monosaccharides which are absorbed as they are released from the food matrix in conjunction with water, vitamins and minerals. On average a 20 kg dog absorbs about 3 litres of fluid daily, 50% of this volume is absorbed in the jejunum, nearly 40% in the ileum and 10% in the colon. During absorption the lumen is cleared of virtually all ingested nutrients plus more than 90% of ingested sodium, potassium and chloride. On a macro scale the large intestine is involved only in a 'mopping up' operation. In certain disease states the small intestine can secrete over 4 litres of fluid daily into the lumen—initially causing diarrhoea by overloading the absorptive capacity of the terminal small intestine and colon and if unchecked leading rapidly to dehydration and subsequently death.

The chyme is mixed with more enzymes in the duodenum, some originate from the duodenal mucosa others from the pancreas. Although a single organ, the pancreas has two discrete functions as an exocrine gland (secreting enzymes into the gut) and an endocrine gland (secreting hormones into the blood, e.g. insulin). It also secretes large volumes of bicarbonate salts into the duodenum optimising luminal pH for pancreatic and intestinal enzymes. The array of pancreatic enzymes include inactive proteases, lipases (fat digesting) and amylases (carbohydrate digesting). Intestinal enzymes, at the luminal brush border, generally catalyse the latter stages of digestion.

Two hormones, secretin and pancreozymin, regulate pancreatic output. Both are produced by cells of the intestinal mucosa and are released into the bloodstream under the influence of specific conditions. Acidity in the small intestine promotes release of secretin which stimulates the pancreas to increase bicarbonate secretion. In contrast pancreozymin release is provoked by the presence of partially digested food in the small intestine, and stimulates the release of enzyme-rich juices.

Digestion and Absorption in the Small Intestine

Carbohydrate

Glucose is the most common dietary unit of carbohydrate. It is however usually present as a component of more complex carbohydrates which must first be degraded to their constituent sugars by both luminal enzymes secreted from the pancreas and those integral to the intestinal mucosa. Starch is the main dietary carbohydrate, being the predominant form of stored energy in most foods of plant origin; its molecular weight varies between 100,000 to over a million which highlights the degree to which it must be enzymically degraded prior to absorption of its individual subunits. Sucrose (glucose–fructose) and lactose (glucose–galactose) are the other main dietary carbohydrates. Sucrose occurs naturally in sugar cane and beet and to a lesser extent in fruits and some roots such as carrots. Lactose occurs only in milk and is less common in the diets of adult cats and dogs (see Chapter 4).

Starch is a polymer of glucose molecules comprising two main fractions: amylose, a straight chain polymer of repeating glucose units, and amylopectin which comprises about 75% of dietary starch and has the same basic structure as amylose but also contains branches in the molecule approximately every 25 glucose residues along the chain.

Starch is initially broken down by the enzyme α-amylase. However because it has little specificity for the branched links, α-limit dextrins (branched oligosaccharides) comprise 20–35% of the final products of intraluminal starch hydrolysis before absorption from the small intestine (Gray, 1975). The remainder is composed of the disaccharide maltose and the trisaccharide maltotriose yielded from the first stage of starch digestion.

The α-dextrins have an average molecular weight of around 1800 and their final digestion along with maltose and maltotriose is facilitated by specific enzymes (oligosaccharidases) integral to the small intestinal brush border membrane which cleave the larger sugars in to transportable monosaccharides. By the time it reaches the terminal duodenum most starch has been degraded to manageable oligosaccharides, commonly containing 3 to 9 glucose units.

Four brush border enzymes are significant in the final stage of digestion of the main dietary carbohydrates prior to absorption, maltase (glucoamylase), sucrase, iso-maltase (α-dextrinase) and lactase (β-galactosidase).

Maltase sequentially removes straight-linked glucose residues from oligosaccharides with 2–9 residues—it is about 20 times less specific, however, for the branched links in α-dextrins. Sucrase plays a dual role by hydrolysing maltose and maltotriose products of starch to free glucose and also by cleaving sucrose to glucose and fructose. Iso-maltase is physically linked to sucrase in the intestinal membrane but acts completely independently to cleave straight and branched links of α-dextrins. This enzyme appears to be the only intestinal oligosaccharidase with significant specificity for branched link of the molecule.

In summary, degradation of α-dextrins (the most complex unit yielded by α-amylase degradation of starch in the lumen) can involve up to 3 brush border enzymes sequentially removing single glucose residues. Maltase and iso-maltase have high specificity for removal of the first or second residue (Gray et al., 1979). When the branched unit becomes terminal, iso-maltase is the enzyme with the highest specificity to cleave the glucose stub to yield maltotriose. Both sucrase and maltase are then capable of degrading maltotriose and maltose to glucose—sucrase apparently having the highest specificity in this regard.

Lactase is primarily responsible for splitting the 'milk sugar' lactose to its component glucose and galactose units, its activity is greatest in young animals and it is commonly absent or less active in adults. For this reason a maximum of 1 g of lactose/kg bodyweight/ day is recommended for adult dogs and cats (see Chapter 4).

Differences in carbohydrate digestion the dog and cat

Pancreatic amylase activity is approximately three times higher in the dog than the cat. High levels of dietary starch can lead to a six-fold increase in amylase activity in canine small intestinal chyme compared with a two-fold increase in the cat (Meyer and Kienzle, 1991). This adaptation can take up to two weeks after the introduction of a new diet to dogs but may take months in cats. Brush border enzymes also tend to have lower activities in the cat. That cats can only tolerate a dietary starch level of about 4 g/kg body-

weight per day before the onset of diarrhoea whilst dogs can tolerate up to 2.5 times this amount (Kienzle, 1989), providing it is well cooked, highlights the practical importance of these differences in the nutrition of these two species.

Absorption of Monosaccharides

The lipophilic brush border of the small intestine is relatively impermeable to the monosaccharides. Entry of these sugars into enterocytes (cells of the intestinal mucosa) is, therefore, dependent on an active transport mechanism in which energy is required to pump the sugars across the membrane. This process is substrate specific and is restricted to those monosaccharides released from normal dietary sugars. Other monosaccharides of a similar size, like mannose or fucose for example, have no transport mechanism and remain unabsorbed in the lumen of the healthy small intestine—and if present in sufficient quantities may cause an osmotic diarrhoea. Glucose and galactose, on the other hand, have almost identical structures and use the same specific transport mechanism—often termed the glucose carrier (Kimmich and Randles, 1980). Glucose or galactose readily bind to a specific transport protein on the membrane surface, this is presumed to be a hydrophobic protein (known as a symport) with a water filled channel through which the monosaccharide moves into the enterocyte (Fig. 3.6). The process is driven by a sodium gradient which is maintained by metabolic energy. Sodium binds to the same receptor as the sugar and both enter the cell simultaneously; the monosaccharide is effectively 'pulled' across the brush border membrane as the Na^+ moves down the gradient. The rate and extent of glucose transport depends on the concentration of this gradient.

In addition to this specific carrier mechanism, a second major monosaccharide mechanism appears to be responsible for the transport of dietary fructose produced by sucrose hydrolysis. This carrier is still poorly understood, but it is known to operate at an efficiency of about 50–75% of that of the

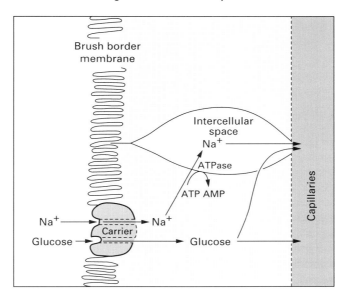

FIG 3.6: A schematic representation of carrier-mediated glucose transport in the intestine. Two molecules of Na$^+$ bind per molecule of glucose (or galactose) and the driving force is provided by (Na–K)-ATPase in the laterobasal membrane. (From Gray, 1975.)

glucose transporter. It is, however, totally independent of glucose transport and is probably also sodium independent (Fridhandler and Quaster, 1955). Digestion of dietary carbohydrate is an efficient process, with one notable exception. Even individuals with optimal lactase activity exhibit comparatively slow hydrolysis of lactose—the process does not provide enough glucose and galactose to saturate the final transport mechanism, into the enterocyte (Gray and Santiago, 1966). Thus, unlike the other dietary oligosaccharides for which the transport step appears to be rate limiting in the overall assimilation, the rate limiting step for lactose is the brush border hydrolysis. Lactase deficiency is common amongst adults of many species, including cats and dogs (see earlier section). In addition, the relatively slow rate of lactose hydrolysis can have implications for individuals with intestinal dysfunction. It has been shown that generalised intestinal dysfunction can depress sucrose and maltose hydrolysis by 40–50%, whilst lactose hydrolysis is depressed by over 75% (Gray *et al.*, 1978). Because this is the rate limiting step in lactose assimilation the implications are more severe. Furthermore in addition to its relative sensitivity to intestinal injury, lactase often remains depressed for longer after intestinal insult than sucrase or maltase.

The stages of carbohydrate digestion and absorption are summarised in Fig. 3.7.

In response to the level of carbohydrate in their diet dogs are able to regulate the rate of small intestinal absorption of monosaccharides. In contrast, the sugar transporters in the feline small intestine are unresponsive to varying levels of dietary carbohydrate (Buddington *et al.*, 1991). This further explains the relatively low dietary threshold cats have for carbohydrate, but in an evolutionary sense their inability to regulate fits with the normally low carbohydrate intake of cats as strict carnivores.

Protein

Pepsin activity in the stomach initiates the digestion of dietary protein. However, the contribution of the gastric phase to overall protein digestion is mainly cleavage of macromolecular proteins to smaller polypeptides and accounts for less than 10% of total protein digestion. The small intestine is the major

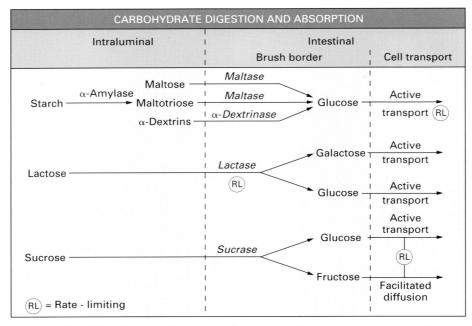

FIG 3.7: Digestion of carbohydrates by brush-border enzymes and mode of absorption of monosaccharides into the epithelial cell.

location for protein digestion—both exogenous (i.e. dietary) and endogenous (i.e. from intestinal secretions) proteins being cleaved to small peptides and free amino acids. Endogenous proteins include secretions of the oral cavity (saliva), stomach, intestine, liver (bile) and pancreas—and can account for up to 50% of the total protein digested especially when dietary protein is low (but not necessarily inadequate). The array of digestive enzymes and cells desquamated during normal turnover of intestinal mucosa constitute most of the endogenous protein fraction.

As in the stomach, intestinal proteases are also secreted as inactive proenzymes by the pancreas, as shown below. The presence of bile stimulates intestinal mucosa to release the brush border enzyme, enterokinase—which activates trypsinogen to trypsin by cleaving a hexapeptide. Trypsin in turn activates more of itself and other pancreatic proenzymes to yield an array of activated proteases.

The endopeptidases trypsin, chymotrypsin and elastase split the internal peptide linkages of larger proteins to smaller peptides. The

three enzymes differ in their substrate specificity. Trypsin, for example can only hydrolyse a peptide chain at a lysine or arginine residue, whilst exopeptidases like carboxypeptidase A + B cleave terminal amino acids from these peptides.

The end result of the action of these intraluminal enzymes is some free amino acids but mainly small peptides (Kim and Erickson, 1985). Free amino acids are absorbed directly by enterocytes, the larger peptides undergo further digestion by brush border aminopeptidases. Short peptides, mainly di- and tripeptides, are readily absorbed into enterocytes where they are cleaved by intracellular peptidase. Amino acid absorption is an active

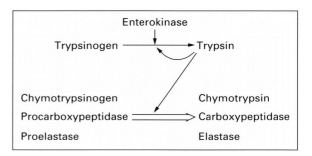

process closely linked to sodium transport—similar in principle to that described earlier for monosaccharides. At least nine distinct transport systems have been identified for amino acid movement across the luminal brush border membrane—each characterised by a specific carrier protein (Hopfer, 1987). The presence of a number of specific amino acid transporters is highlighted by the occurrence of certain genetic defects of amino acid absorption in which failure to absorb either several individual amino acids (e.g. Hartnup's disease—trypsin and other amino acids) or a specific amino acid (such as methionine) is observed (Alphers, 1987). It has been established recently that small peptides, primarily di- and tripeptides are transported intact across the brush border membrane by carriers distinct from single amino acid transporters. There are two significant implications arising from this observation. It is thought that the transport mechanism is dependent on co-transport of hydrogen ions rather than sodium ions as described for amino acids. This is the first observed example of this process in a mammalian system. Secondly, it is unclear how much dietary protein is transported as peptides, but in view of the satisfactory nutritional status of patients with Hartnup's disease, the fraction is probably substantial.

In young animals intact proteins are absorbed across the intestinal mucosa, a process known to transmit passive immunity in many mammalian species by the transport of maternal immunoglobulin. This process occurs by an enfolding action of the enterocyte membrane, known as pinocytosis, in which the protein is engulfed by the cell. Some passage is also thought to occur between adjacent cells. Whilst this phenomenon is of little nutritional significance with regard to the overall uptake of dietary nitrogen, it may be significant in the development of food allergies and even some other less well defined diseases.

Transport of amino acids across the inner membrane of the mucosal cell into portal blood occurs primarily by a sodium independent mechanism with broad specificity for neutral amino acids. A sodium-dependent mechanism does operate at this membrane, but this appears to ensure an adequate supply of amino acids to the enterocytes; for their *own* metabolism in the absence of an adequate luminal supply (Hopfer, 1987).

Glutamine is the amino acid most extensively metabolised by intestinal tissue, with about 25% of the total plasma glutamine being metabolised during each pass through the mucosal tissue bed (Windmeuller, 1982). This is of particular significance in disease where intestinal mucosa is damaged, and glutamine is required at an optimal level to expedite regeneration of intestinal mucosa and enterocyte function. Diets with high glutamine levels may therefore be beneficial in this condition.

Fat

Fat absorption is a multistep process requiring the co-ordination of three separate stages, luminal, mucosal and secretory. (i.e. lymphatic or portal venous transport). Most dietary fat is digested as triglycerides, hydrolysis is initiated in the stomach by the action of lingual lipase that operates at gastric pH without the need for bile salt emulsification. When these free fatty acids leave the stomach they are implicated in the stimulation of cholecystokinin–pancreozymin (CCK–PZ) release by the duodenal mucosa.

Pancreatic lipase in the proximal duodenum is responsible for the lipolysis of the bulk of dietary fat. Fatty acids are released from the individual triglycerides yielding a range of hydrophobic products—monoglycerides, diglycerides and free fatty acids (Fig. 3.8). This process is facilitated by a second pancreatic secretion, co-lipase, a low molecular weight (10,000) protein which binds to the surface of bile salts and lipids, thereby creating an interface at which hydrophilic lipase and hydrophobic lipids can interact.

Optimal lipolytic activity depends on the presence of pancreatic lipase, co-lipase and bile salts, and the co-ordination of their secretion with the presence of lipid substrate in the proximal duodenum. CCK–PZ, released in response to the presence of fat and protein in the duodenal lumen, ensures this coordination

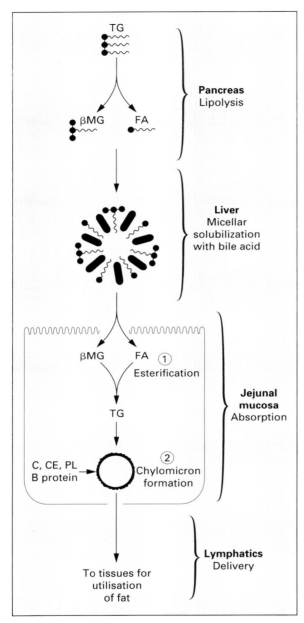

FIG 3.8: A schematic representation of intestinal fat absorption showing the participation of the pancreas, liver and intestinal mucosal cells.

chyme stimulates secretion of pancreatic fluid and bicarbonate, raising the luminal pH to 6–6.5 and creating an optimal environment for lipolytic activity.

The hydrophobic nature of the products of triglyceride lipolysis and other lipids like cholesterol and fat-soluble vitamins requires that they are emulsified with bile salts in mixed aggregates or micelles to permit efficient absorption. The micellar complex transports lipids across the so-called 'unstirred layer' which covers the surface of the intestine and brings them into intimate contact with the microvillus membrane.

Bile acids synthesised in the liver are conjugated with the amino acids taurine or glycerine to yield tauro- or glyco-bile salts; the cat, unlike most mammals conjugates bile acids only with taurine (the implications of the cats peculiar taurine metabolism have been discussed in the previous chapter). Bile salts are lost daily in the faeces and an equivalent amount is synthesised in order to maintain the total bile salt pool. The process of enterohepatic circulation, whereby bile salts are actively reabsorbed form the terminal ileum, enables the liver to secrete a volume of bile 10–15 times that of the body pool daily into the small intestine. Salts may be recycled several times during the course of a single meal.

Enterohepatic circulation illustrates the obligatory requirement for active absorption of bile salts in the ileum. Whilst hepatic synthesis of bile salts can compensate for modest ileal malabsorption (through disease or resection), more complete malfunction causes faecal losses of bile salts to exceed the hepatic capacity for re-synthesis, resulting in reduced concentrations of jejunal bile and leading to steatorrhoea (excessive fat in the faeces).

The passage of fatty acids and monoglycerides across the microvillus membrane is a passive process since the lipids diffuse from the micelles into the lipid-rich layer of the epithelial cell membrane. Inside the cell the lipids become associated with a binding protein which is thought to transport them to the endoplasmic reticulum, the site of triglyceride re-synthesis in the cell. Transport of fatty

by stimulating the pancreatic secretion of lipase and co-lipase and causing the simultaneous contraction of the gall bladder enabling secretion of bile salts. The horse, which has no gall bladder, produces bile salts as they are needed. Furthermore, the low pH of gastric acids away from the site of absorption helps also to maintain a concentration gradient for continued passive uptake.

In order to facilitate transport of absorbed lipids, triglycerides and cholesterol are incorporated into lipoproteins. This involves the formation of triglyceride-rich lipoproteins (chylomicrons) within the endoplasmic reticulum and Golgi apparatus of intestinal epithelial cells; the process is similar to the production of secretory proteins by cells throughout the body. The resulting chylomicron, which contains more than 84% triglyceride and a small amount of cholesterol encased in a hydrophilic film of phospholipid and protein, is discharged from the inner cell membrane into the lymphatic system.

Unlike sugars and amino acids which are transported to the liver in portal blood, lymphatic circulation is required in order for the chylomicrons to reach the systemic circulation. It is worth noting at this stage that fatty acids containing 12 or less carbons (i.e. medium and short chain fatty acids) are not incorporated into chylomicrons but pass directly into the portal blood, and in fact are absorbed directly from the lumen without the need for micellarization. Neither of these fat types are normal dietary components, but some forms of medium chain triglycerides (e.g. coconut oil) can be used to provide energy supplementation under circumstances where normal fat digestion is impaired; for example exocrine pancreatic insufficiency (EPI) or lymphangectasia (congestion of the lymph ducts).

Minerals and Vitamins

The ash fraction of a diet represents the total mineral content and generally speaking it forms about four or five percent of the dry matter of a normal dog or cat diet. Minerals are usually absorbed in an ionised form. The means of absorption varies slightly according to the site in the intestine; for example in the jejunum sodium uptake is linked to the active uptake of glucose, whereas in the ileum and large intestine it is an entirely active process. The absorption of minerals is dependent

on the body's requirement for that mineral, but is also dependent upon its form in the diet.

For example the absorption of calcium is controlled by a number of factors as outlined below: Calcium absorption is directly related to the amount that is soluble and ionised in the small intestine. Many calcium salts, such as phosphate and carbonate are insoluble at neutral pH and digestion of foods at the acid pH of the stomach increases the amount of soluble calcium. Amino acids such as arginine, lysine and tryptophan enhance calcium absorption, as do the antibiotics penicillin, neomycin and chloramphenicol. Calcium absorption is depressed by corticosteroids, thyroid hormone, oestrogens, excessive unabsorbed fatty acids, diets rich in oxalates or phytates (from cereals), phosphorus or excessive antacid administration. The nature of the calcium transport mechanisms are not well understood. In the duodenum calcium is transported by a carrier-mediated energy-dependent mechanism which is regulated by hormones and vitamin D. Since this mechanism is located only in the duodenum, through which chyme passes fairly rapidly, it is the limiting factor in the overall absorption of calcium, which has been reported to be in the region of 50–80% of that consumed in the dog (see *Nutrient Requirements of Dogs*, 1985).

Iron is another interesting mineral in terms of its absorption. It is absorbed mainly in the duodenum and proximal jejunum. In general only 50% of dietary iron is released from food during digestion, the percentage being less for cereals and vegetables and greater for animal-derived foodstuffs. After digestion, the absorption remains optimal as long as the iron remains in solution, with the ferrous form (Fe^{2+}) of iron being better absorbed than the ferric form (Fe^{3+}). In practice iron deficiency is rare in normal circumstances and dietary iron is readily available. It forms chelation complexes with large mucopolysaccharides secreted in the stomach and also with ascorbate, citrate, some hexoses and certain amino acids, which maintain iron in a soluble form.

The fat soluble vitamins (A, D, E and K) are absorbed with dietary lipids, and no problems with absorption will be seen unless fat malabsorption occurs. Water-soluble vitamins are absorbed by a process of simple diffusion, with several exceptions. For example, vitamin B_{12} can only be absorbed after binding to a protein known as intrinsic factor which is produced by the gastric mucosa, whereas folic acid is absorbed by an active process in upper part of the small intestine.

The Large Intestine

Structure

Digesta passes from the small to the large intestine through the ileo-caecal valve. The large intestine is relatively short in the dog and cat, since its principle purposes are to absorb salt and water, in contrast to the large colon of herbivores, required to digest polysaccharides (Table 3.1). The large intestine consists of the caecum, colon and rectum. The caecum forms a blind-ended pouch below the junction of the small and large intestines at the ileo-caecal valve. The caecum is a vestigial organ in the cat, of moderate size in the dog and voluminous and sacculated in the horse. The colon which makes up most of the large intestine is not coiled like the small intestine but consists of three relatively straight portions—the ascending colon, the transverse colon and the descending colon, which extends to the pelvic inlet where the rectum begins. The large intestine has no villi, and its surface is flat. Straight tubular glands, the crypts of Lieberkühn, extend from the surface through the mucosa. These glands primarily contain mucous cells in their deepest portions and both mucous and epithelial cells near the surface. The mucus secreted by these glands is alkaline (bicarbonate) and its function is to protect the large intestine mucosa from mechanical and chemical injury. The mucus provides lubrication to facilitate passage of the faeces, whereas the bicarbonate ions neutralise irritating acids produced by local bacterial fermentation.

Movement

Most of the time, movements of the large intestine are slow and nonpropulsive. This slow wave activity occurs at less than five cycles per minute and shuffles contents in a back-and-forth mixing movement that exposes the colonic contents to the absorptive mucosa. Occasionally colonic contents are seen to move rapidly over a relatively long distance; this is due to marked and simultaneous contraction of large segments of the ascending and transverse colon, driving the contents one-third to three-quarters of the length of the colon in a few seconds. These contractions, known as mass movements, drive the colonic contents to the distal portion of the large intestine where material is stored until defaecation occurs. Mass movements are initiated by food entering the stomach (the gastrocolic reflex), which is mediated from the stomach to the colon by gastrin and the extrinsic autonomic nerves, often leading to the urge to defaecate.

When mass movements of the colon move faecal material into the rectum, the resultant distention of stretch receptors in the rectal wall initiates the defaecation reflex. This causes the internal anal sphincter (which is composed of smooth muscle) to relax and the rectum to contract. If the external anal sphincter (which is composed of skeletal muscle) is also relaxed then defaecation occurs.

Absorption

The surface area of the large intestine is limited, and although it is capable of absorption of water and some electrolytes, it has none of the transport mechanisms needed for the absorption of organic nutrients. The ileo-caecal flow of water is high—particularly in the horse, notably due to the low dry matter percentage of ileal chyme (only 5% compared with around 20% in the dog and cat). Water is absorbed by being drawn into the intracellular spaces by concentration gradients produced by the absorption of electrolytes and volatile fatty acids. In its simplest form, the large intestine serves merely for transporta-

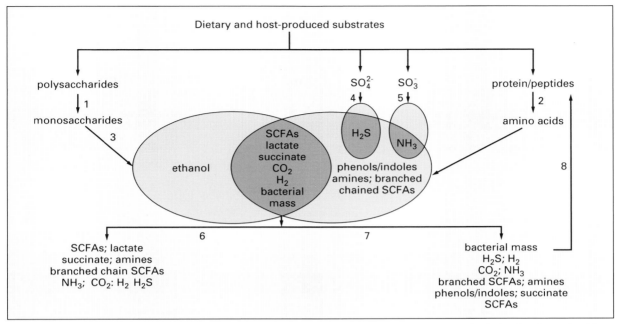

FIG 3.9: Summary of the degradative processes mediated by intestinal bacteria. (1) Hydrolysis of polysaccharides by bacterial enzymes; (2) hydrolysis of proteins and peptides by bacterial enzymes; (3) fermentation of sugars; (4) sulphate reduction; (5) nitrate reduction; (6) host absorption or metabolism; (7) excretion in breath, faeces and urine; and (8) death/breakdown of bacteria releases cell contents. Adapted from: Cummings and Macfarlane, 1991. *J. appl. Bact.* **70**, 443–459.

tion of undigested food constituents and for the recovery of endogenously secreted substances (water, electrolytes). In some species, e.g. the horse, this organ has differentiated, increased in size and has become very important in the digestion of food.

Bacterial Fermentation

The microflora of the gastrointestinal tract consists of hundreds of different species of bacteria. With the exception of ruminants, most bacteria in the gastrointestinal tract are concentrated in the large intestine (around 10^{10}–10^{11} per gram of digesta). The numbers and distribution of the intestinal flora are established soon after birth and are normally stable, maintained in a symbiotic relationship with the animal. Approximately 99% of the bacteria in the normal, healthy intestine are anaerobic, and the main bacterial species present in a healthy dog or cat are Streptococci, Lactobacilli, *Bacteroides* and *Clostrid-*

ium. Diet can have an effect on the bacterial content of the gut—with Lactobacillus being higher in young animals fed a milk diet and Clostridium higher in dogs and cats fed an all meat diet.

The organic matter flowing into the large intestine is comprised mainly of dietary fibre, plus some undigested carbohydrate (starch) and proteins. The bacterial colonies which reside in the large intestine are able to partially digest this organic material, to produce the short chain fatty acids acetate, butyrate and propionate, as shown in Fig. 3.9. These can then be absorbed to provide an energy source for the animal. Fermentation also leads to the production of gases such as carbon dioxide, hydrogen and ammonia. The degree of fermentation in the large intestine will be dependent on the composition of the diet, the size of the large intestine, the residence time of the foodstuff and the bacteria present. The residence time of undigested food in the large intestine of the dog and cat is around 12 hours, compared with 36 hours in the horse.

It is predicted that 25% of the total digestion of food occurs in the large intestine of horses compared with around 8% in dogs (Drochner and Meyer, 1991). On average 4–8 g dry matter per kg bodyweight reaches the large intestine per day in various species—the levels being higher in the horse and lower in the dog. Approximately 75–80% of this dry matter is organic. The shorter large intestine of the dog is capable of fermenting fibre to only a limited extent around 7–35%, compared with 36–68% in the horse, which has a much more developed large intestine (Drochner and Meyer, 1991). The digestion of starch in the large intestine of the horse is virtually complete, whereas in the dog it ranges from 15–100% depending on the type and form of the starch. The amount and composition of protein entering the large intestine varies considerably. It is difficult to quantify protein digestion in the large intestine since nitrogen is present as undigested protein, mucus, bacteria, urea and ammonia. The undigested food proteins are essentially cell wall proteins, with the bacterial proteins making up about 20–30% of large intestinal protein content. The amount of fat entering the large intestine is low in most species, either because the fat content of the diet is low (as in the horse) or because precaecal fat digestion is very effective (as in the dog and cat). The presence of fat in the large intestine can inhibit bacterial fermentation. Large intestinal chyme also contains a considerable amount of ash, which is comprised of high levels of sodium and chloride ions as well as unabsorbed minerals such as calcium and phosphorus.

Constipation and Diarrhoea

Any residues left undigested, together with water, minerals and bacteria are stored in the rectum until evacuation. Defaecation is usually under voluntary control, involving relaxation of the anal sphincter, but diarrhoea or illness may override this control. Diarrhoea is defined as the frequent evacuation of watery faeces. It can be caused by factors which either increase the rate of transit of chyme through the intestine so that digestion and absorption cannot occur and/or factors which impair the absorption of water in the large intestine. Persistent diarrhoea (and vomiting) can be fatal due to the loss of inorganic ions and the effects of dehydration, although occasional vomiting and diarrhoea can be caused by nothing more serious than a rapid change in diet or a bout of overeating.

If defaecation is delayed too long, constipation may result. When the colonic contents are retained for longer periods of time than normal, then more water is absorbed and the faeces becomes hard and dry. Decreased colon motility following certain drug treatments or as a result of aging can also lead to constipation, as can a low-fibre diet.

Dietary fibre is essential in the diet of the horse (Chapter 8). Furthermore, a certain level of fibre in the diet of the dog and cat can help reduce the incidence of both diarrhoea and constipation. It tends to normalise transit time through the gastro-intestinal tract.

Flatulence

Occasionally, instead of faecal material passing from the anus, intestinal gas or flatus is passed. This gas is derived both from swallowed air and bacterial fermentation in the large intestine. The quantity and nature of the flatus is dependent on the type of food eaten and the nature of the colonic bacteria, but typically contains nitrogen, carbon dioxide, hydrogen, and much smaller amounts of methane, hydrogen sulphide and ammonia.

Water Balance

Water is often neglected as a nutritional requirement because of its ready availability in most temperate climates. However, the requirement for water is at least as important as that for other nutrients; life may continue for weeks in the complete absence of food, but only for a few days or even hours when water is not available.

Water fulfils many roles within the body. It

is an excellent solvent, and this property makes possible all the complex chemistry of cellular metabolism. As the principle constituent of blood, water provides a vital transport medium, taking oxygen and nutrients to the tissues and removing carbon dioxide and metabolites. Blood also carries antibodies and white cells which protect the body from disease.

Water contributes to temperature regulation in several different ways. Firstly, the blood transports heat away from working organs and tissues, thereby preventing dangerous temperature increases. Then, by redirection of some of the blood through superficial veins, heat can be transferred to the skin and lost to the environment through radiation, convection and conduction. Heat loss may be further increased by the evaporation of water from the skin.

Water is also essential for digestion. Hydrolysis, the splitting of compounds by water, is the means by which digestion occurs. Digestive enzymes are secreted in solution, the better to disperse amongst the foodstuffs. Even the elimination of toxic metabolics via the kidney requires water as a medium. These represent only a few of the many functions of water.

There are several different fluid compartments in the body, which can be grouped together as intra- or extra-cellular fluid (ICF and ECF). ICF represents approximately 50% of the animal's total bodyweight and includes the water inside all cells from red blood cells to the neurones in the spinal cord. ECF is found bathing the tissues in between the cells, and in the blood and lymph. Movement of fluid between these compartments is continuous, different concentrations of electrolytes being maintained by the activity of cell membranes.

Water Output

Water leaves the body by several routes. In normal healthy animals these include losses in expired air, faeces, urine and sweat. These pathways will be discussed separately. In sick animals water loss may be increased markedly through haemorrhage (bleeding), vomiting and diarrhoea. Lactation is another instance of increased loss.

Faeces

The water content of faeces is usually very low compared with the enormous volumes of fluid secreted into the digestive tract, with enzymes, mucus and various electrolytes. The intestines have very efficient mechanisms for water reabsorption and it is only when these are disturbed and faeces evacuated as diarrhoea that this route makes a significant contribution to water loss. The water content of well-formed faeces is in the range of 65–75%, with diarrhoea having a water content of 80–90%.

Evaporative Losses

The uptake of oxygen from inspired air is made by close association between the epithelium of the lung and an extensive capillary network.

However, this also facilitates the transfer of water by diffusion and evaporation into the cavity of the lung, and the water is then lost in expired air. This 'respiratory water loss' is unavoidable. In hot weather evaporation is an important temperature regulating mechanism because of the body heat used to vaporise the water. Horses lose heat by conventional sweating, whereas dogs pant and hang out their tongues, and cats cover their coats with saliva by repeatedly licking. In extreme conditions there may also be some slight evaporation through the foot pads in dogs and cats. Although these mechanisms aid temperature control, they may increase water loss.

Urine

The kidney is the only organ in the body which can control water loss. In addition to this it also regulates acid–base balance and the concentration of many electrolytes. Most animals have two kidneys situated in the abdominal cavity, one either side of, but ventral to (below, or in front of) the spinal column. The blood supply is provided by the renal artery and vein.

The kidney consists of a network of thousands of tubules. Each tubule has a blind end, or 'glomerular capsule' which envelopes a knot of capillary blood vessels known as the glomerulus. There is a wide difference in pressure between the capillary and the capsule and this differential causes continuous movement of small molecules and fluid into the capsule from the capillary. Large molecules, such as proteins and the various blood cells, cannot pass into the tubule unless there has been damage to the glomerular or tubular walls. Indeed, one of the indications of kidney damage is the presence of proteins in the urine. In healthy animals therefore, the fluid enters the tubule as an 'ultrafiltrate' of blood, and the rate of entry depends upon the difference of pressures in the two systems.

CHAPTER 4

A Balanced Diet for Dogs and Cats

KAY E. EARLE and PHILIP M. SMITH

Introduction

Cats and dogs are the most common companion animals kept in the home environment. Although domesticated for thousands of years, the dog and cat still show the dietary traits of their carnivorous ancestors. For many years they were considered to be very similar in their dietary needs but it has become apparent that there are fundamental differences in their nutritional requirements. Many of the differences between these animals can be traced back to a consideration of such diverse details as chromosome numbers (Felidae, have either 36 or 38 and Canids range from 38 to 78), dental and cranial anatomy (cats have 30 permanent teeth compared with 42 in dogs), smaller intestinal to body length ratio (4:1 in cats and 6:1 in dogs) and a much smaller caecal tissue weight in proportion to body weight in cats. Much of this evidence is consistent with the diet of dogs containing more plant material than the cat's. In fact the dog can be considered to be an omnivore rather than an obligate carnivore like the cat.

The key differences between the nutritional requirements of dogs and cats are shown in Appendix I and were discussed in greater detail in Chapter 2. Suffice it to say that the standard daily ration for these two species would be different in analytical profile to account for these differences in nutritional requirements.

The provision of a balanced diet for our cats and dogs must inevitably include consideration of a number of factors which are all interlinked and should not be discussed in isolation. These include nutrient content, energy content, digestibility and palatability of the food. *A balanced diet can be defined as a diet which allows no net gain or loss of nutrients from the body to maintain a state of metabolic equilibrium.* This diet will supply all the key nutrients needed to meet the daily needs of the recipient animal together with the quantity of energy required to sustain the animal's life stage. The role of a balanced diet is to promote a long and healthy life for a dog or cat that will keep the animals in peak condition and hence reduce their susceptibility to disease.

The major dietary components of protein, fat and carbohydrate provide the animal with a source of energy following their breakdown in the gut and subsequent absorption. Protein and carbohydrate release a similar amount of energy per unit weight whilst fat liberates twice as much; all three nutrients can be interchanged in the diet to provide the required

TYPICAL NUTRIENT CONTENT OF PREPARED FOODS FOR DOGS AND CATS									
Food type	Moisture %	Protein %	Fat %	Ash %	CHO %	Ca %	P %	ME MJ/100g	Protein energy %
Dogs									
Wet food									
Meat in jelly	83	7.0	4.5	1.5	4.0	0.3	0.2	0.32	37
Meaty chunks in jelly	79	7.0	4.5	2.5	7.0	0.5	0.4	0.36	32
Meat in jelly for puppies	80	9.0	6.0	2.5	2.5	0.4	0.3	0.39	39
Semi-moist food	21	17.0	8.0	8.0	45.0	1.2	1.1	1.22	23
Dry food									
Complete	6	30.0	11.0	7.0	46.0	1.3	1.1	1.47	34
Mixer biscuit	6	13.0	7.5	7.0	66.5	1.3	0.5	1.44	15
Mixture									
Canned meaty dog food and biscuit (3:1 by weight)	63	10.0	6.0	3.0	18.0	0.6	0.5	0.63	23
Cats									
Wet food									
Meat in jelly	83	9.0	5.0	2.5	0.5	0.5	0.4	0.29	52
Meaty chunks in jelly	80	7.5	3.5	2.5	6.5	0.5	0.4	0.34	38
Meat in jelly for kittens	80	11.0	5.5	2.0	1.5	0.3	0.2	0.36	52
Dry food									
Complete	7	33.0	6.5	7.0	46.5	0.9	1.1	1.30	43

TABLE 4.1: Typical nutrient content of prepared foods for dogs and cats.

amount of energy, however the over-riding factor will be dependent on the composition of the raw materials used. Table 4.1 shows proximate analyses of prepared pet foods typically fed to dogs and cats and their overall energy content per 100 g wet weight. The fundamental requirement of all species is for energy which provides the power for all cells to function. If this need is coupled to the requirement for an individual nutrient then providing the correct quantity of energy also maintains a nutrient balance. When all key nutrients are balanced to the energy content of the food then animals requiring a higher plane of nutrition e.g. during gestation and lactation, will consume more energy and this will inevitably lead to a higher intake of all the key nutrients. The composition of the diet can be adjusted for any specific life stage and hence satisfy an increased demand for a specific nutrient.

There are four major food groups (meat and fish, dairy and eggs, cereal and vegetables and fats and oils), which are able to provide a variety of the dietary nutrients in differing quantities. The quality of any dietary ingredient is determined not only by its nutrient profile but also its digestibility. For example, the protein value of the food groups can be ranked in terms of their bioavailability as dairy/egg > meat/fish > vegetable/cereal.

Meat and Fish

Meat consists of muscle tissue from animals, birds, or fish, often with intramuscular fat, connective tissue and blood vessels. Depending on the site within the body this material may vary quite widely in its amino acid profile and digestibility, this will ultimately affect its nutritive value. In this respect pre-cooking of the material before feeding may be of benefit. There are also real differences in nutrient content between meat from different parts of the body, particularly in the fat content (Table 4.2). A very useful source of information on the nutrient content of meat and other foods is found in the 1991 revision of McCance and Widdowson's *The Composition of Foods*. The protein quality of meat from the sources mentioned above is relatively high, however, the fat from different animal sources may differ in degree of saturation. Raw materials from

TYPICAL NUTRIENT CONTENT OF SOME MEATS AND MEAT BY-PRODUCTS							
	Water g/100g	Protein g/100g	Fat g/100g	Calcium g/100g	Phosphorus g/100g	Potassium g/100g	Energy MJ/100g
Raw lean meats							
Pork	71.5	20.6	7.1	0.008	0.20	0.09	0.62
Beef	74.0	20.3	4.6	0.007	0.18	0.15	0.52
Lamb	70.1	20.8	8.8	0.007	0.19	0.10	0.68
Veal	74.9	21.1	2.7	0.008	0.26	0.36	0.46
Chicken	74.4	20.6	4.3	0.01	0.20	0.32	0.51
Duck	75.0	19.7	4.8	0.012	0.20	0.29	0.51
Turkey	75.5	21.9	2.2	0.008	0.19	0.30	0.45
Rabbit	74.6	21.9	4.0	0.022	0.22	0.36	0.52
Offal							
Udders	72.4	11.0	15.3	0.26	0.24	0.13	0.76
Fatty lungs	73.1	17.2	5.0	0.01	0.19	0.15	0.48
Sheep lungs	76.0	16.9	3.2	0.01	0.20	0.19	0.40
Brains	79.4	10.3	7.6	0.01	0.34	0.40	0.46
Stomachs (pig)	79.1	11.6	8.7	0.03	0.11	0.14	0.52
Spleen	75.9	17.0	6.5	0.03	0.22	0.40	0.52
Kidney (beef)	79.8	15.7	2.6	0.02	0.25	0.23	0.36
Heart	70.1	14.3	15.5	0.02	0.18	0.32	0.83
Heart (trimmed)	76.3	18.9	3.6	0.005	0.23	0.28	0.45
Liver (fresh)	68.6	21.1	7.8	0.001	0.36	0.40	0.68
Green tripe	76.2	12.3	11.6	0.01	0.10	0.12	0.65
Dressed tripe	88.0	9.0	3.0	0.08	0.04	0.08	0.26

TABLE 4.2: Typical nutrient content of some meats and meat by-products.

poultry and pigs have a lower degree of saturation compared with beef and lamb but this will have little effect on overall digestibility.

Offal meats such as liver, kidney, tripe, melts and lights contribute very different nutrient contents depending on the food provided for the donor animal. Meats in general are devoid of free carbohydrate since in any host species this is quickly converted to energy sources by the host on ingestion and the energy reserves are in fat deposits. Meat materials are usually low in calcium, compared with phosphorus, which can lead to under-mineralisation if fed as a sole source of food; supplementation is necessary prior to feeding. Most meats and some offals are deficient in vitamins A and D, however the liver and to a lesser degree the kidney are good sources of these vitamins. Liver can contain so much vitamin A that over-use in diets can lead to ingestion of levels causing serious hypervitaminosis A and illness—this is discussed in more detail later in the chapter. Apart from this, meats are a good source of protein supplying most essential amino acids, fats, iron and some B group vitamins. Meat materials are highly palatable to cats and dogs and when correctly supplemented they make excellent food. Of the other meat by-products surplus to the human food industry, animal carcasses contain a high proportion of bone and consequently are good calcium and phosphorus sources. Added in the correct amounts they can help to offset or complement the deficiencies of the 'muscle meats'. Handling difficulties may be a limiting factor in the home but these materials are often used in the manufacture of prepared pet foods.

Fish are commonly divided into fatty fish and white fish. White fish like cod, haddock, plaice, whiting and sole usually contain less than 2% fat, whereas the oily or fatty fish like herring, mackerel, pilchards, sardines, sprats, tuna, salmon, trout and eels may have very much more, between 5 and 18%, depending on the season or stage of maturity of the fish when caught. In general, white fish are very similar to lean meat in composition. The protein is of similarly high quality and the vitamins A and D are generally absent or present only in trace amounts. But fish muscle does contain adequate amounts of iodine and because bones are frequently consumed with the flesh of fish, the calcium and phosphorus content is much better balanced. Filletted fish with bones removed is seriously deficient in calcium and phosphorus. The flesh of oily fish does contain vitamins A and D and the livers of fish like cod and halibut are particularly rich sources of these fat-soluble vitamins. Whole fish, including the bones (if made safe by cooking or grinding up) are better balanced sources of nutrients for dogs and cats than most meats.

Fish are usually less palatable than meats but nevertheless are generally quite well accepted by these animals, but their smell and appearance may be less acceptable to some dog and cat owners. Like meat, fish can contain parasites and should be cooked before being used as food. In addition, some fish muscle contains the enzyme, thiaminase, which breaks down the vitamin thiamin. This enzyme is destroyed or inactivated by heat and provides another reason for cooking fish before feeding it to pets.

Dairy Produce and Eggs

Items of dairy produce contain higher quality protein with a more comprehensive amino acid profile than meat or fish; they are usually more palatable to dogs than cats but may be well liked by both species. A small number of cats and dogs may be unable to tolerate more than a minimum intake of milk sugar, lactose, which can result in diarrhoea in those animals who have insufficient lactase, the digestive enzyme which breaks down lactose into its component parts. These animals are easily identified and should not be given milk or other dairy products. For other animals about 20 ml milk per kg body weight is a useful guideline. Milk contains most of the nutrients needed by cats and dogs but is a poor source of iron and vitamin D. It is a good source of readily available energy, protein of high quality, fat, carbohydrate, calcium, phosphorus and several trace elements, vitamin A and B complex vitamins. The riboflavin is sensitive to sunlight and most of it will be destroyed

together with vitamin C (not a dietary essential for cats and dogs) if it is exposed to sunlight for more than an hour or two. Whole pasteurised milk contains per 100 ml around 275 kJ of energy, 3.3 g protein, 3.8 g fat, 4.7 g lactose, 0.12 g calcium and 0.095 g phosphorus. Skimmed milk has almost no fat or vitamins A, D and E and a higher concentration of protein and lactose than whole milk. Milk from Channel Island breeds of cow contains rather more fat and protein than the average and has an energy content of about 320 kJ/100 ml. Otherwise it has similar properties. Goats' milk is very similar to cows' milk in composition and is of no extra value to cats and dogs.

Cream, which is rich in fat and fat-soluble vitamins and cheese (a coagulation of milk protein) are very palatable to cats. Most of the protein, fat, calcium and vitamin A content of milk is retained in the cheese whilst the milk sugar and B vitamins are removed in the whey. Most cheeses have similar amounts of protein and fat except for those like cottage cheese which are made from skimmed milk and so contain almost no fat. Cream cheeses contain the greatest amount of fat. Dairy foods provide an excellent source of many nutrients to our cats and dogs; however, if the individuals are lactose intolerant these foods should be avoided.

Eggs are often fed to growing dogs and cats because they contain a rich source of iron, protein, riboflavin, folic acid, vitamin B_{12} and vitamins A and D. They also contain appreciable amounts of most other nutrients except vitamin C and carbohydrates. The white of the egg is almost all protein and water with trace minerals and some B vitamins (except niacin which is found in trace quantities in eggs). Most of the B vitamins and all of the fat soluble ones are found in the yolk which contains more fat and protein and much less water than the white. Raw egg white contains an anti-nutritional factor (avidin) which affects the bioavailability of the vitamin, biotin. Heating destroys this biotin-binding effect and increases the digestibility of the egg white, therefore it is advisable to cook eggs before feeding.

Cereals and Vegetables

Cereals form the next most important food source and they include grains such as wheat, barley, oats, rice, rye, maize and some sorghums. Cereal grains consist of the germ or embryo surrounded by a starchy endosperm whose function is to provide storage carbohydrate (starch) and some protein (gluten) to support the growth of the germ. This endosperm is itself surrounded by an aleurone layer, a thin layer of cells rich in protein and phosphorus outside which is the tough outer seed coat. Milling of cereals separates the various layers to give the bran which contains the tough outer coat rich in polysaccharides, cellulose and hemicelluloses (dietary fibre) and the flour which is largely composed of starch and gluten from the endosperm. The whole grains of the common cereals wheat, oats, barley, rice, maize, contain about 12% moisture, 9–14% protein, 2–5% fat and about 70–80% carbohydrate as starch. Wheat, oats and barley have a higher protein content and less fat than maize and rice.

Generally cereals are used as a source of energy for dogs and cats but they also provide a significant proportion of the protein to the total diet. There is little to choose between the cereals in the quality of their protein, however they do contain substantial amounts of other nutrients, particularly thiamin and niacin. The bran portion obtained by separating the seed coat is a good source of dietary fibre and of phosphorus, but unless cooked before feeding, much of the phosphorus will be unavailable because it is present in a complex called inositol phosphate or phytate. Cooking improves the availability of the phosphorus. The high fibre content makes bran useful as a bulking agent when diets of low nutrient content are required but depending on the extraction rate of the milling process, bran will supply dietary energy and can contain up to 800 kJ/100 g. Bran is also useful in that its inclusion in the diet up to a certain amount tends to have a good effect on faeces consistency, reducing the likelihood of constipation or diarrhoea. Wheat germ or other cereal germs are rich sources of thiamin, protein, fat and vitamin E.

Tapioca is considered to be a cereal by many people, although in fact it is made from the starchy root of the cassava plant. This has much less protein than rice and is almost entirely starch with small amounts of minerals and only traces of vitamins. For practical purposes it should be treated as contributing only energy.

Cereals are not particularly palatable to cats and dogs, even when moistened with water and usually need to be fed as only part of the diet. As a general rule cats should receive less than 5 g digestible carbohydrate per kilogram of body weight and dogs 12 g otherwise digestibility and faeces quality may be poor. The nutrient content is of lower digestibility than many other foods particularly if the cereals are not finely ground or cooked. Fine grinding or cooking markedly increases the digestible energy and digestible dry matter values. This is mainly because of the effect in improving the digestion of starch.

Green (or stem) vegetables are generally not very well accepted by dogs and cats and their bulk and indigestible fibre content means that they would need to eat large amounts to obtain a significant contribution to their nutrient intake. Vegetables are a good source of the B vitamins but these may be destroyed during cooking and lost in the cooking liquor if it is discarded. Dogs, but not cats, may obtain some of the dietary precursor of vitamin A from vegetable sources in the form of β-carotene which is present in some roots and tubers. Root vegetables are rich in starch and are poorly digested in their raw state, by cats and dogs. Cooking gelatinises the starch and makes it more digestible; most dogs will eat cooked potatoes and carrots. The third type of vegetables includes the pulses e.g. peas and beans, which are relatively rich in protein and provide more energy than stem or root vegetables, other than potatoes. They are fairly good sources of most B vitamins. Green peas, broad beans and runner beans are acceptable to most dogs when cooked but rarely form a major part of their diet. Soya beans are a major source of protein and energy for humans in many parts of the world and are used world-wide for animal feeding, either as whole beans or more commonly after processing to remove the oil. Soya beans have a tough outer seed coat or hull which is removed mechanically before extracting the oil by a combination of grinding and treating with solvents. The residue which is left contains the protein, carbohydrate and mineral portions of the bean with some small amounts of oil. It may be 'toasted' or heated to inactivate certain anti-nutritive factors (trypsin inhibitors and haemagglutonins) contained in the seed. These substances are sensitive to heat and the amount of heat treatment is controlled so that it does not at the same time denature the soya protein and reduce its nutritive value. The heat-treated or toasted de-fatted soya bean meal usually has a protein content of 48–50%, 30% carbohydrate, mainly as sugars not starch, 1–2% fat, about 5–15% minerals with 3–5% crude fibre. The protein obtained from soya bean meal is of good quality and has high levels of the essential amino acids. De-fatted soya bean meal may be used directly as a food ingredient or it may be further processed to make textured vegetable protein, often abbreviated to TVP. Most legumes contain complex carbohydrates and simpler sugars which are resistant to digestion by the digestive enzymes of animals like the dog, cat or man. They pass undigested into the large intestine where they may be fermented by bacteria with the consequent production of flatus. This problem of flatulence is common when feeding legumes like green peas and beans. The degree to which flatulence occurs in cats and dogs fed soya products or peas and beans seems to depend on the amounts fed and on the susceptibility of the animal, which in turn probably depends on the bacterial flora present in the gut. The production of flatus should not be overlooked when considering the use of these materials in foods for dogs and cats.

Fats and Oils

Fats and oils include materials like butter, margarine, lard and dripping, the visible fat in meats and the invisible fats in substances like nuts, lean meats and other foods. Oils are distinguished from fats only by their melting point; they are liquid at room temperature, whereas fats are solid. Fats and oils may be further classified as containing largely saturated or unsaturated fatty acids. Most fats contain both kinds of fatty acids but the proportions differ. The unsaturated fatty acids—linoleic and linolenic acids (with 2 and 3 double bonds respectively)—are known as essential fatty acids (EFA) because they are required in small quantities for optimal health and cannot be synthesised by the cat or the dog from other fats. Linoleic acid is widely distributed in vegetable seed oils and occurs in small amounts in some animal fats, particularly pork and chicken fat. Arachidonic acid is an additional fatty acid that is essential to the cat and occurs only in small amounts in some animal tissue fats. It is found predominantly in the phospholipids of the organs (e.g. liver and kidney) of grain-fed animals; small quantities are also present in egg yolk.

All fats yield a similar amount of energy, about 2.25 times that obtained from protein or carbohydrate and their nutritional value in other respects depends largely on their origin and vitamin content. Oil seeds contain 20–40% fat and this is almost pure fat with only traces of minerals and no vitamins apart from vitamin E. Wheatgerm oil is a good source of vitamin E activity followed by sunflower and cottonseed oils. Vegetable fats are usually better sources of unsaturated fats like linoleic acid than animal fats; sunflower oil, soya bean oil and corn or maize oil have about 50% of this fatty acid, safflowerseed oil contains even more, 65–70%, but coconut oil contains hardly any and olive oil only about 10%. Milk fat and beef tallow contain similarly low levels of 2–4%. Fish oils are a good source of fatty acids containing over 20% linoleic acid. They usually contain many other unsaturated fats as well. Most animal fats contain only trace amounts of B vitamins but cod liver oil, butter and margarine are good sources of vitamin A. Margarine and cod

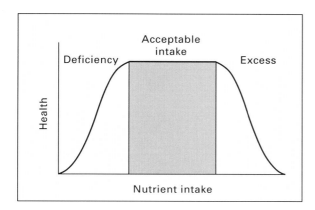

FIG 4.1: Nutrient intake and its relationship to nutritional adequacy

liver oil, or other fish oils like halibut liver oil, are also good sources of vitamin D. Margarine has no inherent vitamins from its component vegetable fats but in many countries must be fortified with vitamins A and D. The very high levels of vitamins A and D in fish liver oil make this suitable for only limited inclusion in the total diet.

Fats, particularly animal fats, are greatly liked by cats and dogs. They add flavour and palatability to foods, help to bind together cereal meals and are therefore more than just a source of energy, EFA and vitamins A, D and E in the diet. Fats are normally highly digestible and slow down the rate of stomach emptying. They are also thought to help give animals a feeling of satiety after meals. Cooking fat which has been repeatedly used for deep fat frying should be avoided as it is likely to contain peroxides and other toxic materials which can be harmful if fed to cats and dogs.

Nutrient Balance

The raw materials described above offer a wide variety of ingredients for use by the pet owner and pet food manufacturers when formulating a balanced diet for a cat or dog. Dietary formulations must always ensure that the food contains more than the minimum

dietary requirements in order to reduce the risk of clinical signs of deficiency. The guidelines provided by the National Research Council in 1985 (Dogs) and 1986 (Cats) on the minimum nutrient requirements are largely based on studies using semi-purified diets which are assumed to be 100% available for metabolism. The food products provided by pet food manufacturers are formulated from raw materials of varying nutrient content and digestibility reflecting the profile of their constituent ingredients. Therefore it is necessary to take these factors into account when assessing the absolute level of a particular nutrient in the food. Individual nutrients should not be considered in isolation as their biochemical interactions are very complex. Figure 4.1 is a schematic diagram of the effect an individual nutrient can have on the health of the animal. The minimum dietary requirement has been clearly defined for most of the nutrients and clinical signs of deficiency have been identified if this level is not reached in the diet. Similarly the maximum tolerable levels of certain nutrients are known and, if the dietary concentration exceeds this limit, toxicity will result. The extent of the plateau between the minimum requirement and the maximum tolerable level will change depending on the individual nutrient and the overall dietary composition. If we take as an example potassium, this nutrient is essential for the maintenance of cellular integrity and in nerve and muscle cell excitability. Low dietary potassium will induce a clinical condition which is characterised by weight loss, lethargy, depressed reflexes and hypokalaemia (low blood potassium levels). The NRC 1986 dietary guidelines for cats suggest a minimum requirement of 4 mg/g dry matter (DM) or 0.19 mg/kJ, based on the feeding of a semi-purified diet. If we investigate the potassium content of meat materials and cereals commonly fed to dogs and cats (Table 4.2) the recorded values differ widely between food groups. If these raw materials are used to provide a high protein diet for a cat it has been observed that the dietary potassium requirement increases to around 0.25 mg/kJ to avoid clinical signs of deficiency and hypokalaemia (Buffington *et al.*, 1991).

This illustrates the difference in measurable quantities of potassium that will satisfy the metabolic need of the cat when fed diets of varying composition. Currently there are no data available on the maximum tolerable level of this nutrient as any excess is eliminated via the urine.

The setting of a maximum tolerable level for an individual nutrient becomes important when chronic feeding has led to either excess storage or acute toxicity and death. An example of a nutrient where chronic toxicity may be a problem if the maximum tolerable dose is exceeded is vitamin A. A number of researchers have shown a severe degree of pathological disturbance characterised by bony exostoses (bony outgrowths) of the vertebrae and long bones with excess dietary vitamin A. If this vitamin is fed in very large amounts there is a gradual progression from excess storage (lipid infiltration of the hepatic cells) to toxicity. Raw liver was observed to be more toxic than retinyl palmitate as it raised blood retinol levels more excessively. The NRC in 1987 documented the maximum tolerable level of vitamin A in dog and cat foods as 1,000 IU/kg DM however studies with cats have shown that this can be exceeded with no adverse effect. A vitamin A intake of 50,000 IU/kg BW, caused excessive liver storage when fed to kittens for 41 weeks but no overt clinical signs (Seawright *et al.*, 1967). This intake is equivalent to about 1,330,000 IU/kg DM in the diet and is well in excess of the NRC (1987) maximum recommendation.

When formulating a complete and balanced diet it would be relatively easy to overlook one factor such as nutrient interaction and provide an imbalance of nutrients unless one's knowledge of nutrition was extensive. Most people who have a cat or a dog do not have the degree of nutritional expertise, access to the appropriate raw materials, the time or the desire to pursue complicated preparation and cooking of food for their pets. As a consequence the manufacture of foods specially prepared for cats and dogs has developed, particularly over the last forty years, into what is now a large and sophisticated industry.

Prepared Pet Foods

Prepared foods are available in many different forms but may be conveniently classified by their water content and method of preservation as indicated in Table 4.3. Prepared pet foods may also be categorised on their nutrient content, i.e. if they are sold as a complete food, a complementary food, a mixer biscuit, a snack product or a treat. Complete foods contain all the nutrients required by the dog or cat for that particular stage of the lifecycle for which they are sold, e.g. for adults, for puppies. They require no supplementation, except that clean fresh water should always be made available. Complementary foods are not intended for use as the only food in a diet. They may be rich in some nutrients but inadequate in others. Mixer biscuits, as their name implies, are to be fed with other ingredients in the diet, either prepared or fresh foods. They are high in energy and are good sources of certain minerals and vitamins but this will depend on the recipe used by the manufacturer. Their major role in the diet is to provide a relatively cheap source of energy and as such they usually complement high-protein canned meats. Because the nutritional needs of dogs and cats are different, manufacturers make foods specifically designed for each species. In practice, it is quite safe to feed cat food to dogs, but it is not safe to feed cats on dog foods and the practice is not recommended. There is adequate choice of brands and varieties to satisfy the needs of most pets within the species-specific range.

Moist Foods

The most familiar types are the canned pet foods which have been available for many years as a variety of meat- and fish-based products or meat, fish and cereal products. These have now been joined by similar products in plastic pots or aluminium trays. Their properties are well known and they are a very reliable, safe and convenient way of providing moist attractive foods which are highly palatable to the cat or dog. The most palatable are those which contain little or no cereal or

CLASSIFICATION OF PREPARED PET FOODS BY WATER CONTENT AND METHOD OF PRESERVATION		
Food type	**Water content %**	**Preservation technology**
Dry	5–12	Drying
Semi-moist	15–50	Reduced water activity by use of humectants, mould inhibitors, low pH
Canned	72–85	Heat sterilisation
Frozen	60–80	Freezing
Chub	70–85	Heat treatment and/or preservatives
Plastic pot	75–85	Heat sterilisation
Aluminium tray	75–85	Heat sterilisation

TABLE 4.3: Classification of prepared pet food, by water content and the method of preservation.

carbohydrate source and are presented as meaty or fishy chunks in gravy or jelly. Those which contain significant amounts of cereals are best described as loaf products, meat and cereal or fish and cereal foods.

The digestibility of these foods for dogs and cats is very good and hence the availability of nutrients is high. Their soft-moist, meaty or fishy nature ensures good palatability. It is possible to feed cats and dogs satisfactorily on these products alone, but because the energy content is relatively low, large amounts are needed, especially for large breeds of dog. This is rather wasteful of protein and even though most canned dog foods are complete diets they are also designed to be highly palatable sources of good quality proteins, vitamins and minerals to be fed in conjunction with cheaper biscuits or other mixers which primarily supply energy with some minerals and vitamins. Feeding recommendations of canned foods which are expected to be fed in this manner usually indicate that the proportions of canned food to biscuit mixer can be varied quite widely to provide a more or less palatable energy-dense mixture to suit the needs of particular feeding situations. For most large adult dogs an equal volume mixture of canned food and reputable mixer biscuits will provide a highly palatable, nutritious meal.

Heat-sterilised moist foods are safe products with a very long storage life, not requiring special storage conditions. They are usually produced by chopping and mixing the main ingredients, adding gravy and processing in a sealed container. The processing involves combinations of time, temperature and pressure of steam which vary with can size and heat transfer characteristics of the recipe, but which are sufficient to kill even the most resistant and harmful bacteria. There is little damage to, or loss of, nutrients from the food except for thiamin which is particularly sensitive to heat and so compensatory amounts are added to maintain adequate post-process levels.

Semi-Moist Foods

Meaty types of dog and cat foods with water contents between 15 and 30% can be preserved with a shelf-life of several months by maintaining a reduced water activity. Water activity is a measure of the water which is available for bacterial or fungal growth in or on the surface of a food. These organisms cannot grow and cause spoilage in dry products (up to 12% moisture) because there is not enough water present. Water activity (A_Δ) is measured as relative humidity at equilibrium with the surrounding environment; most bacteria will not grow at levels below 0.83 and yeasts and moulds below 0.6. The low water activity in semi-moist foods is achieved by the inclusion in the recipes of humectants such as sugars, salt, or glycerol which 'tie-up' the water. Further protection is provided by the use of preservatives such as sorbates to prevent yeast and mould growth or by the reduction of pH (increased acidity) with organic acids. These foods can be made with a variety of ingredients including meat, meat by-products, soya or other vegetable-protein concentrates, cereals, fats and sugars. The technology allows the water content to vary over a wide range and so the product form may be as a fairly dry material (15% water) not dissimilar to a dry food or a soft-moist substance similar in appearance to minced or cubed meat (25–30% water). The most popular forms currently available contain about 20% water and so have a fairly high nutrient density. They are of average or above average digestibility, 80–85% for most nutrients. They do not usually have a strong odour, do not dry up rapidly if exposed to the atmosphere and so can be left in the feeding bowl without becoming unattractive to pet or to owner. Cat products have not been so successful as those designed for dogs and this may be because the cat is more selective in its choice of foods.

Dry Foods

Dry foods are available for both cats and dogs. Dry dog foods are sold as baked biscuits, extruded and expanded biscuits, or as mixtures of meals and flakes. They may be complete foods or formulated as mixers intended for feeding as part of the diet with protein-rich foods such as fresh meats, fish or canned dog foods.

Mixers are usually based on cereals and contain very little, if any, protein concentrates. They are supplemented with minerals and vitamins to provide a complete balanced diet when fed in appropriate amounts with cooked or canned meats. Many of the cheaper kinds are not supplemented and so would require other foods and supplements besides meat to provide an adequate diet. Such mixers are little more than cooked cereal with fat added sufficient for baking or extruding. Good quality mixers are supplemented with extra calcium, phosphorus, trace minerals and vitamins to balance their energy content. When mixed with good quality canned meats they provide adequate amounts of all nutrients.

Complete dry foods for dogs come in similar physical forms but differ in their ingredient content. They are usually formulated to supply adequate amounts of all known nutrients for the stage of life for which they are intended. Loss of nutrients, particularly of vitamins, is limited because the baking or extrusion processes do not require excessive temperatures or time and sufficient supplements are added to counterbalance processing and storage losses. Because they are dry and do not contain enough water for bacterial or fungal growth, they have a long shelf-life and will keep in dry, cool storage conditions for many months. Complete dry foods are made from cereals and cereal by-products; protein concentrates of animal or vegetable origin like, for example, meat and bone meal, fish meal or soya bean meal; fats and mineral and vitamin supplements.

Dry cat foods are available as extruded, expanded biscuits formulated to provide a complete diet for cats of all ages or for adult maintenance. The ingredients of dry cat foods are similar to those of dry dog foods but more emphasis has to be given to the inclusion of proteins and fats of animal origin and some even include fresh meat rather than meat meals. Cats have a need for greater levels of protein than dogs and so protein levels in dry cat foods are frequently higher than in dry dog foods.

Dry foods contain a greater concentration of nutrients and energy per unit weight than foods of higher moisture content and so relatively small amounts are needed to provide a particular quantity of nutrients. Unless they contain large amounts of fibre, the digestibility of dry foods is good but often lower than that of meats and canned foods. Dry foods are considerably better digested by dogs than cats. Baked or extruded biscuits have been partially or wholly cooked and so are a good source of energy for both dogs and cats. They are easy to store and to dispense. The main disadvantage of dry foods is that they are less palatable than moist foods like meat or canned foods. There are considerable variations in palatability between brands of food because manufacturers take considerable trouble to enhance the acceptance of their own particular products. The mixer biscuits for dogs are meant to be fed with canned meat or meat/gravy and therefore the palatability is less of a problem than with complete dry foods. Good quality dry foods for dogs and cats are well accepted and can be used as the sole source of nutrition. Dry foods provide a relatively cheap source of energy which can add considerable flexibility to feeding regimes.

The Right Food

There are now a number of options available to the discerning pet owner when choosing an appropriate feeding regime for their companion animal. Feeding dogs and cats at any stage of their lives is straightforward, using a wide selection of commercially prepared foods. These products provide the pet owner with the reassurance that the feeding regimes they adopt will be nutritionally sound as well as satisfying their other expectations of palatability, convenience, price and suitability for their particular domestic arrangements. Careful observation of the animal's appearance and behaviour will enable the owner to identify the correct level of feeding for their animal, its likes and dislikes and to arrive at a suitable feeding regimen. Finally, the role of a balanced diet is to promote a long and healthy life by providing the individuals with not only an adequate intake of nutrients but a choice of food type and variety that will maintain an adequate intake throughout all stages of life.

CHAPTER 5

Feeding Dogs and Cats for Life

VÉRONIQUE LEGRAND-DEFRETIN and HELEN S. MUNDAY

Introduction

We have already seen, that we can make recommendations on the nutrient level in a food for a group of animals which takes into account the individual variation of those animals. However, during certain periods of the animals' lives the demands made upon the food become much greater, for example during the rearing of young and through the growth period of those young animals. In this chapter consideration will be made of how the nutrient requirements of the animal change through its life and how we may best meet these requirements. Not only do we need to take into account the changes in nutritional requirements but also how our husbandry techniques change, for example number of feeds per day and the format in which we offer food, for instance soaking dry foods for young animals.

Animals eat to obtain energy and we must ensure that the nutrient content is balanced to this energy level. Additionally the palatability of the diet should be such that the animal is happy to take in the amount of food needed. In this chapter we will consider the following groups of animals—gestating and lactating bitches and queens, growing puppies and kittens, senior dogs and cats and, finally, working dogs.

Gestating/Lactating Animals

Ensuring the best possible start in life for young animals begins with the correct nutrition of the pregnant and lactating bitch or queen. In this section we consider how the nutritional requirements of the reproducing animal differ from that of the adult dog or cat at maintenance.

The Breeding Bitch

To plan a sensible feeding programme for the bitch, it is necessary to understand what are the extra nutritional and physical demands made by breeding. Foetal growth of puppies shows that most of the foetal weight gain occurs in the last third of the gestation period. Accordingly, the bitches' energy requirement will not increase until this time although there is considerable development of mammary and uterine tissues beforehand. Overfeeding early on in pregnancy may lead to the deposition of unwanted fat and may predispose to problems at whelping. A satisfactory regime to follow would be to increase the amount of food by 15% each week from the fifth week of gestation, so that at the time of whelping (9 weeks) the bitch will be eating 60% more than when it was mated. It may happen that a bitch with a large litter may have such an

57

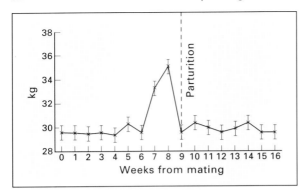

FIG 5.1: Weight change of Labrador Retriever during gestation/lactation (n = 17).

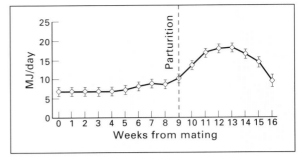

FIG 5.2: Energy intake of Labrador Retriever during gestation/lactation (n = 17).

enlarged abdomen and such reduced activity that its appetite falls during the last week or ten days of pregnancy. In these cases it is sensible to feed several smaller meals during the day. The objective is to have a bitch at parturition which is not overfat and which has maintained its appetite.

Lactation presents the biggest test of nutritional adequacy of any feeding regime. The bitch must eat, digest, absorb and use very large amounts of nutrients to produce sufficient milk of adequate composition to support the growth and development of several puppies. The extra energy needed is dependent of the normal energy intake of the bitch and the size and age of its litter. Consider a Labrador Retriever bitch of 28 kg with a litter of six 4-week-old puppies of 2.5 kg each. At this stage each puppy will require an energy intake of about 2100 kJ/day which is obtained from the bitch's milk. The bitch therefore has to supply 12600 kJ as milk each day. Bitches' milk contains about 5650 kJ/litre and so the amount of milk needed is at least 2.3 litres. There are obviously some losses of energy in the production of milk by the bitch but if it is assumed that the process has an efficiency of 75% then in order to produce 12600 kJ as milk, the bitch must obtain 12600/0.75 or 16800 kJ from its food. In addition, to maintain its own body weight and condition it will need its usual 6400 kJ/day. Its total energy requirement is therefore 23200 kJ or nearly four times its maintenance requirement. Obviously, it is strongly recommended to feed

such an amount of food in several small meals of a highly palatable and digestible diet. Figures 5.1 and 5.2 show the bodyweight changes and the average energy intakes of a group of Labrador Retriever bitches during gestation and lactation studies at the Waltham Centre for Pet Nutrition (WCPN). If the bitch is unable to produce enough milk or to eat the amount of food it needs, then early supplementary feeding of puppies may be necessary if they are to do as well as they should.

The above calculation is based on estimates of the energy needs for satisfactory milk production but requirements for other nutrients are similarly increased. Protein quantity and quality will affect milk production. It is therefore necessary to ensure that the extra food supplied is of good quality and is not made up only of high fat or high carbohydrate foods. In 1981 Romsos et al. suggested that gestating/lactating bitches needed carbohydrate in order to whelp and rear healthy puppies. However, a later study conducted on Labrador Retriever and Beagle bitches (Blaza et al., 1989) showed no differences in the reproductive performance of bitches fed either a diet in which there was no available carbohydrate or one where carbohydrate represented 11% of the estimated metabolisable energy. This is probably due to a higher protein content of the diet fed in the Blaza et al. study which supplied sufficient gluconeogenic amino acids to allow the maintenance of the plasma glucose level. It was then concluded that, while carbohydrate is physiologically essential, it is not an indispensable component of the diet of gestating/lactating bitches.

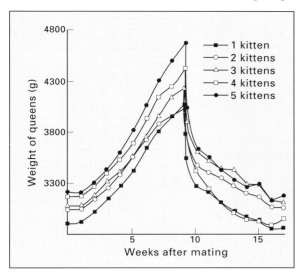

FIG 5.3: Weight changes during gestation and lactation in queens with different sizes of litter. Data show the means for 15 queens in each group. (From Loveridge and Rivers, 1989.)

Finally, the breeding bitch does not require special vitamin/mineral supplements if a balanced diet is used.

The Breeding Queen

Unlike the dog, once the queen has been successfully mated her food intake begins to increase almost immediately. Similarly, the changes in body weight occur steadily and gradually almost from the first day of pregnancy, which is quite unusual amongst the mammalian species. Loveridge and Rivers (1989) comprehensively reviewed this subject and noted that the overall mean weight gain during pregnancy (irrespective of the size of the prospective litter) was 39% of the pre-mating weight. However, the weight gain did vary with the size of litter and this could be explained though the application of the following linear equation:

$$\text{Weight gain } (g) = 888.9 + 106.5N$$

where N = number in litter.

Interestingly, this constant figure of 106.5 is remarkably close to the mean birth weight of kittens (106.2 g) in the colony of cats studied (at WCPN).

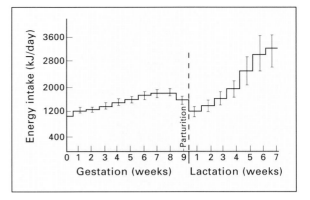

FIG 5.4: Variation in food intake of cats during gestation and lactation. Data are for 10 cats fed singly. Their patterns of weight change parallel those observed in the main group of animals. Food intakes after the fourth week of lactation probably include significant amounts of food eaten by kittens. (From Loveridge, 1986.)

These changes in weight gain during pregnancy (and lactation) are illustrated in Fig. 5.3.

Loveridge and Rivers hypothesised that this unusual pattern of body weight gains in the cat is the result of deposition of extra uterine tissue early on in pregnancy. The later weight gains become increasingly attributable to the foetus itself as pregnancy progresses. The normal gestation length in the domestic short-haired cat is in the region of 64 days i.e. very similar to that of the dog. Recent data from WCPN have shown a mean gestation length of 65.8 ± 2.5 days but with a significant difference between cats of different genotypes. A number of other factors such as litter size and parity number affected the length of gestation; however, there was a great deal of variation both within individuals from one pregnancy to another and between individuals (Munday and Davidson, 1992).

Obviously the queen needs to support this weight gain and thus food and energy intakes increase during pregnancy and, indeed, the rise in energy intakes follow quite closely the changes in body weight. Loveridge (1986) individually measured the energy intakes of ten queens and Fig. 5.4 illustrates the changes in energy intake both during gestation and lactation.

In terms of energy intakes on a body weight

basis, these increase from the adult maintenance requirement of 250–290 kJ/kg BW to over 370 kJ/kg BW during gestation. On a practical note cats rarely overeat and so *ad libitum* feeding is quite acceptable. In this way the queen will be able to take exactly the amount of energy she needs and the owner simply needs to offer slightly more than they would to the non-pregnant queen. The pregnant animal is more susceptible to nutrient deficiencies or excesses and so the diet should be carefully regulated at this time. For example the calcium : phosphorus ratio has to be more tightly controlled as the earliest stages of bone development are occurring in the kittens *in utero* and the minimum protein requirement is higher than for maintenance.

As with the bitch, the life stage of lactation in the queen proves to be the biggest test of nutritional adequacy, for not only is the queen having to obtain nutrients for herself but also for her many kittens through the milk she must provide. At birth the kittens weigh between 85 and 120 g and there can be from 1 to 8 kittens in a litter. These figures will vary with such factors as the breed of cat and adequacy of the diet, but apparently not with the queen's body weight (Loveridge, 1987). For the first four weeks of their life the kittens are entirely dependent on the milk supplied by the queen and so the queen's energy requirements are far greater during lactation than through gestation as, by this time, the kittens are growing very rapidly. Although the kittens begin to take solid food from about four weeks of age the queen's energy requirement remains elevated until after the completion of weaning (when the kittens are about seven to eight weeks of age), as the queen is still suckling (although to a lesser extent) and is also rebuilding her own body reserves. At parturition the queen loses only about 40% of her body weight in the form of the foetuses and afterbirths and during the eight weeks of lactation she gradually reduces her weight until she achieves her pre-mating weight.

The energy requirements of the lactating queen depend on the number of kittens she is rearing and their age, as both these factors will affect the amount of milk she needs to produce. Queens' milk has an energy density of 444 kJ/100 g which is high compared with cows' milk at 272 kJ/100 g (Baines, 1981). The queens' energy intake during lactation changes as illustrated in Table 5.1.

As can be seen from these figures the queen's energy requirements are some three to four times that of her maintenance requirement. Thus the queen needs to be offered a highly palatable and digestible food which is high in energy. The queen will need to take frequent small meals to ingest the required amount of food and, again, *ad libitum* feeding is highly appropriate as the queen will be able to control her energy intakes very successfully. Fresh water should be available at all times as a great deal will be lost from the queen's metabolism through the production of milk. As in the case of the pregnant queen, the nutrient levels in the food need to be more tightly controlled than would be the case for an adult at maintenance. For this reason

DAILY ENERGY INTAKES OF QUEENS DURING LACTATION						
Week of lactation	Litter size					
	1	**2**	**3**	**4**	**5**	**6**
1	248	315	381	447	514	514
2	273	344	414	485	555	555
3	298	389	481	572	663	663
4	323	439	555	671	788	788
5	348	485	622	759	900	1037
6	373	564	754	945	1136	1327

TABLE 5.1: Daily energy intakes of queens during lactation. All figures as kJ/kg BW and are based on the voluntary intakes for queens of mean BW 3.8 kg after parturition. Values from about week 4 include energy intake of kittens. (Data taken from NRC, 1986.)

AVERAGE ANALYSES OF MILK OF VARIOUS SPECIES				
	Bitch	**Cow**	**Goat**	**Cat**
Moisture %	77.2	87.6	87.0	81.5
Dry matter %	22.8	12.4	13.0	18.5
Protein %	8.1	3.3	3.3	8.1
Fat %	9.8	3.8	4.5	5.1
Ash %	4.9	5.3	6.2	3.5
Lactose %	3.5	4.7	4.0	6.9
Calcium %	0.28	0.12	0.13	0.04
Phosphorus %	0.22	0.10	0.11	0.07
Energy* kJ/100g	565	276	293	443

TABLE 5.2: Milk analysis of various species. * Calculated using protein 16.72 kJ/g, fat 37.62 kJ/g and lactose 16.72 kJ. (From Baines, 1981.)

foods which are specially designed for the lactating queen should be fed, as they will have been formulated with these factors in mind. For example the levels of certain vitamins, minerals and protein will be more tightly controlled and the food will have an increased energy density. It is important that if a balanced food is fed no further supplementation of nutrients is made as this may actually cause an imbalance in the food.

Growing Animals

In relation to body weight, the energy and nutrient requirements of growing animals are far greater than those of adult animals. The young growing animal needs a higher plane of nutrition to fuel its rapid growth and to provide the boundless energy which is so characteristic of puppies and kittens. In this section we will discuss the best way to feed growing animals.

The Growing Puppy

The first two weeks of a puppy's life are spent eating and sleeping, their energy needs catered for by their mother. Clearly during this period they have no need for any external source of food. Naturally if the bitch cannot produce enough milk, for example if it is a very big litter, or of course if the puppies are orphaned, then they must be hand-reared.

The most obvious alternative to a bitch rearing her puppies is for another bitch to act as a foster mother. However, the chances of a bitch at the right stage of lactation and with sufficient resource to rear a litter being available at just the right time are poor. It can however, be done and good communications within a breed club, for example, obviously improve the chances of finding a suitable foster mother. Motherless puppies have vital requirements in two main areas: provision of a suitable environment and nutrition. There are two very important aspects of husbandry to consider: the ambient temperature around the puppies and the stimulation of urination and defaecation of each puppy. Ideally the environment would be controlled by means of an incubator. Alternatively a heating pad with adequate insulation of the pen can be used. After puppies have fed, a vital aspect of tending motherless puppies is to simulate the mother's tongue action on the ano-genital area which provokes reflex defaecation and urination. The necessary result can be achieved by applying a piece of damp cotton wool at the ano-genital area or simply by running a dampened forefinger along the abdominal wall. Between 16–21 days puppies no longer require stimulation to urinate and defaecate and from 28 days, when they completely control their body temperature, they begin to explore their surroundings and become more independent. Puppies grow at a rapid rate and will double their birth weight in a matter of days. Because of this, puppies require quite large quantities of their mother's milk or a food which can substitute for it. The food has to be a concentrated source of nutrients based on the composition of normal bitches' milk. Table 5.2 shows the average composition of milk from bitches, cows, goats and queens. It is clear that cow's and goat's milks are inadequate as a substitute for rearing puppies since the protein, fat and calcium levels are too low. Queen's milk would also be inadequate because of lactose and calcium levels. Many commercially available bitch's milk substitutes are now available. They are usually based on cow's milk which has been modified to resemble bitch's milk more closely. They can be administered by means

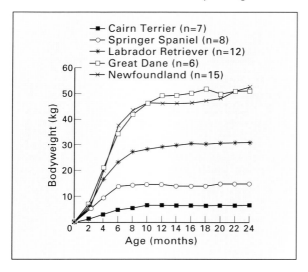

FIG 5.5: Growth curves of different breeds.

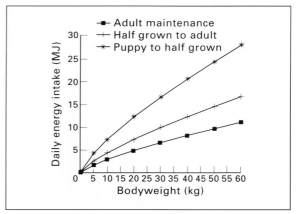

FIG 5.6: Energy requirement of dogs during different life stages.

of a small syringe or a puppy feeding bottle. Dried milk feeds should be reconstituted daily and fed warm (38°C). Food must be given slowly and must not be forced into the puppy. When feeding from a miniature bottle, the hole in the teat may need to be enlarged so the flow is improved and the puppy does not suck in air. When puppies begin to explore their surroundings, a high quality puppy food can be introduced. This can be mixed with the milk substitute to begin with and then offered separately.

In the early stages of weaning (3–4 weeks after parturition) the bitch's milk is still the most important source of nutrients and the puppies' digestive system is learning to handle new sources of nutrients. At this age, puppies should be encouraged to take soft, wet food or even small kibbles of dry food moistened with warm gravy or water. Although many people assume that milk and milky feeds should figure prominently in a weaning regime, they are not essential. Highly palatable, energy and nutrient dense foods are those most suitable for weaning. Puppies become fully weaned between 6 and 8 weeks of age, after which they are ready to leave their mother's side. At this stage, it is better to feed them 4 small meals a day, rather than allow them continuous access to food.

It has already been pointed out that the

dog is unique in that it has the widest range of normal adult body weight (in relative terms) within any single species. The rate of growth in the early stages is very rapid and, in general, most breeds of dog will attain 50% of their mature adult weight at a similar age, that is 5–6 months. However, because of the wide variation in adult body weight, different breeds continue to mature at different relative rates and larger breeds take longer to reach their adult body weight than smaller breeds of dog (Fig. 5.5). The feeding regime is of primary importance in determining the growth rates of large and giant breeds in particular. Individuals of certain of these breeds that have the most rapid growth rates are more prone to disturbances in their skeletal development. Maximum growth is not necessarily compatible with optimum growth and, provided that skeletal and muscle development is not ultimately impaired, then an elongated period of growth is likely to be advantageous to such puppies. At weaning (between 6 and 8 weeks of age), the energy requirements per unit body weight of puppies are about double those of an adult of the same breed. As they grow this requirement decreases to reach progressively the adult requirement. Table 5.3 and Fig. 5.6 illustrate the energy needed by puppies of different sizes as their weight increases. Assuming canned puppy foods are around 500 kJ/100g, a 12 week old puppy weighing 20 kg would have to eat 2.5 kg/day to cover its energy

ENERGY REQUIREMENTS OF PUPPIES					
Bodyweight	**Age in months**				
kg	**2**	**3**	**5/6**	**12**	**24**
1	1045	836	523	523	523
2	1756	1404	1053	878	878
5	4368	3494	2098	1747	1747
10	7348	7348	3528	2940	2940
15		9961	5977	3983	3983
20		12519	7411	4941	4941
30			11721	8373	6696
50			19646	12281	9823
60				14082	11265

TABLE 5.3: Energy requirements of puppies throughout growth (kJ/day).

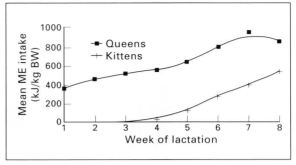

FIG 5.7: The energy intakes of queens and kittens during lactation. Based on a mean of 3 kittens/litter.

requirement. It is easy to understand why this amount of food should be provided in at least four meals a day to allow the digestive system to digest and absorb all the nutrients necessary for optimal growth. The frequency of feeds should then be decreased throughout growth (twice daily after they have reached 50% of adult body weight) until adult body weight is reached. It is not recommended to feed puppies *ad libitum* since they will tend to overeat and could become obese.

Not only do puppies have higher energy requirements than adults but some nutrient requirements are also different. For example, protein and calcium requirements are higher because of the strenuous demands of growth. The level of the latter in a puppy food should be related to the level of phosphorus since the ratio of these two minerals plays an important role in the development of bones. Thus it is important to provide puppies with a food which is energy dense, highly digestible and which takes into consideration their special needs.

The Growing Kitten

As we have seen, during the first few weeks of life the young kitten is entirely dependent on its mother's milk and at this stage no supplementary feeding is required. During this early period, a growth rate of nearly 100 g per week is desirable but obviously there will be a good deal of individual variation determined by factors such as nutrition, breed and the

queen's body weight. Occasionally the milk supply from the queen is inadequate and in this case specially manufactured milk replacers should be given at frequent intervals throughout the day and night. As with puppies, not only assistance with feeding but with urination and defaecation will be required.

From about three to four weeks the kittens become increasingly interested in the solid food which is being offered to their mother. To assist them in taking food it is often useful to offer some finely chopped moist food in a shallow tray or dry food that has been soaked in water or milk. This food may be the same as the queen's food or may be a food specifically designed for kittens. Once the kittens

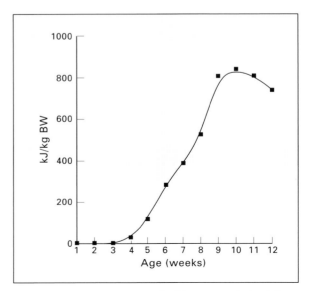

FIG 5.8: Mean daily metabolisable energy derived from solid food in kittens.

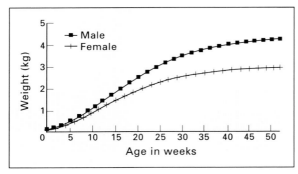

FIG 5.9: Standard growth curve of kittens (males and females).

begin to take solid food the process of weaning has begun and the kittens will gradually take more solid food until they are fully weaned at about seven to eight weeks.

Some recent work at WCPN (Munday and Earle, 1991) has begun to quantify the energy intakes of kittens as derived from solid food before the completion of weaning. At four weeks of age the kittens only eat about 10 g of food per day (10–40 kJ/kg BW) and the greater proportion of their requirements are accounted for by the queen's milk. By five weeks of age (sixth week of lactation) they eat between 15–45 g/kitten/day which equates to 250–350 kJ/kg BW (depending on the energy density of the diet). The kittens' energy intake from solid food increases from zero in weeks two and three of lactation to over 800 kJ/kg BW at eight weeks of age. This means that the kittens' food intake constitutes a considerable proportion of the total energy consumed by queen and kittens during the latter stages of lactation. The mean energy intake of the kittens as a proportion of the total energy ingested by both the queen and her kittens,

rises from 5% in week four of lactation, to 20% and nearly 30% during weeks six and seven respectively, (Figs 5.7 and 5.8).

Once the kittens have been weaned from the queen there is no real need to provide milk. Indeed, as the digestive tract of the kittens develops, their ability to digest the lactose in the milk gradually becomes less until in some adult cats there is complete intolerance. If there is a desire to give milk, special lactose reduced milk drinks are available but fresh water is adequate and should be available at all times.

As young animals, the kittens have a small physical capacity and it is not only advisable to feed energy dense foods but also to feed frequently. Unlike puppies, kittens should ideally be fed *ad libitum* as they are unlikely to overeat. At weaning the kittens should weigh between 600 g and 1000 g. At this stage it is already apparent that the male kittens are heavier than the females and this is a trend which is maintained throughout life. Typical growth curves for kittens of the domestic shorthaired variety are illustrated in Fig. 5.9.

Energy requirements are at a peak (840 kJ/kg BW) at about the age of ten weeks and after this point the energy requirements per unit body weight gradually decrease although they remain relatively high for at least the first six months of life whilst growth is rapid. Table 5.4 illustrates the amount of food needed by a growing kitten when fed a specially designed kitten food containing 370 kJ/100g or a lower energy food (typical of a food for adults) of 290 kJ/100g.

Not only should food for kittens be of higher energy concentration than that fed to adults but some nutrient requirements are

FOOD AND METABOLISABLE ENERGY INTAKES OF GROWING KITTENS					
Age in weeks	8	12	18	25	40
Mean weight (g)	800	1300	2000	2800	3300
Energy requirements (kJ/kg BW)	910	830	580	420	370
Total energy requirement (kJ)	730	1080	1160	1180	1220
Required food intake of 370 kJ/100g product (g)	196	291	313	317	330
Required food intake of 290 kJ/100g product (g)	251	372	400	405	421

TABLE 5.4: Food and metabolisable energy intakes of growing kittens.

also higher and this should also be taken into account. For example, the requirement for dietary protein which is already relatively high in the adult cat is even higher in the growing kitten (about 10%). Calcium and phosphorus levels also have to be maintained within quite tight margins as any excesses or deficiencies could result in bone deformities. It is also important to stress that the addition of calcium supplements to an initially balanced diet will cause as many problems as the feeding of a poorly designed diet. The role of taurine in reproduction and growth is now well documented and all foods for growing kittens should be replete in this amino sulphonic acid.

Most kittens have gained 75% of their ultimate body weight by the time they are six months old and weight gains after this tend to be due to developmental changes rather than the earlier picture of skeletal growth. Thus, after the age of six months it is suitable to feed the young cat on foods formulated for adults rather than kittens. As adult males are significantly heavier than females their growth and development will continue for considerably longer than for a comparable female. For both sexes the food intakes will take some time to reach adult levels as slow growth is still occurring between six to twelve months, but stabilisation should occur by the end of the first year. Times of feeding can also be reduced over the first six months although many people continue to offer multiple meals throughout the day.

Senior Animals

Ageing is unfortunately irreversible. The modifications of cellular and biochemical structures which are responsible for the decrease of the mass and activity of muscular, nervous and other tissues cannot be stopped. However, providing adequate food which takes into account the particular needs of the older animal can help in the care and management of this life stage. As in human medicine, improvements in the prevention of serious life-threatening contagious diseases means that more of our companion animals are reaching old age and so this will obviously become an area of increased interest.

Senior Dogs

Defining when a dog is old is less easy than for the cat because of the variety of breeds. However, it is known that small and medium-sized breeds have, in general, a higher life expectancy than giant breeds. For example, Beagles are not really considered to be old before 10, while Great Danes are elderly when they are 8 years old.

Evidence from human studies shows that the energy expenditure of the elderly is reduced for two main reasons. Firstly, their level of activity falls, although there is a wide variation between individuals. Secondly, there is a decrease in lean body mass and a subsequent decline in their basal metabolic rate. There is little quantitative data available to support this hypothesis in dogs or cats, however, two studies showed a decline in average daily energy intake of dogs (Fig 5.10, Finke, 1991; Table 5.5, Kienzle and Rainbird, 1991) with increasing age. These data suggest that although there are differences between individual dog breeds, the over-riding factor is the age of the dog. Therefore, it is not unreasonable to conclude that older pets should be fed to a lower energy requirement than the younger adult in order to reduce the risk of obesity. However, this does not mean that a senior food should be less energy dense. This is because senior animals may have a poor appetite and will therefore tend to eat less food. For example, it has been reported that 16% of 12-year-old dogs presented at a veterinary hospital were underweight while only 5% were overweight (Kronfeld *et al.*, 1991).

There is little information available from studies assessing the digestive function and the particular nutrient requirements of elderly pets. One study (Sheffy *et al.*, 1985) compared digestive functions of sixteen pure-bred Beagles of 10–12 years old with those of eight one-year-old Beagles. The results showed that the apparent digestibilities for protein, fat, ash and energy were *higher* for old dogs than

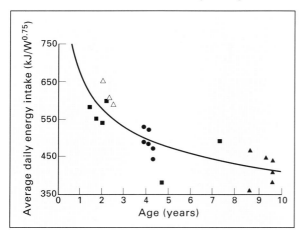

FIG 5.10: Effect of age on the energy intake of older Labrador Retrievers(▲), young male Labrador Retrievers (△), Siberian huskies (■) and Beagles(●). The line represents the best fit regression equation $Y = 665X^{-0.2066}$; $r = 0.812$. (From Finke, 1991.)

for young dogs regardless of the diet being fed. Buffington *et al.* (1989) measured digestibilities in three groups of Beagles aged 2–3, 8–10 and 16–17 years. Mean nutrient digestibilities for dry matter, nitrogen and fat were all slightly lower for the oldest group but none of the differences between the groups were statistically significant. Despite these results, Wannemacher and McCoy (1966) showed that the optimal intake of dietary protein was greater for 12- to 13-year-old Beagles than for 9- to 12-month-old Beagles. Protein requirements in elderly dogs can then be considered from two aspects. On one hand, old animals may have a higher protein requirement than younger adult dogs whilst on the other hand, high levels of protein may be a

INFLUENCE OF AGE ON MAINTENANCE ENERGY REQUIREMENT OF LABRADOR RETRIEVERS	
Age (yr)	kJ digestible energy/kg BW$^{0.75}$
≤ 1	633±28[a] (15)
1-2	593±29[ab] (13)
3-7	572±15[b] (45)
>7	464±19[c] (14)

TABLE 5.5: Relationship between age and energy requirement. Values are means ± SEM; number of observations in parentheses. Age groups that are not marked with the same letter differ significantly. (From Kienzle and Rainbird, 1991.)

stress on renal function. Thus protein requirements should not necessarily be met by simply increasing the level of protein in the diet. It is better to consider the quality of the protein rather than the quantity in itself. For old animals, the protein should have a high biological value and a high precaecal digestibility to reduce the formation of bacterial metabolites.

The tolerance of fat is not generally reduced in elderly dogs and their requirements for essential fatty acids appear to be the same as their younger counterparts.

Very little is known about the mineral requirements of old healthy animals. It can be argued that the intake of certain minerals, such as sodium and phosphorus, should be reduced because they are commonly associated with '*diseases*' of the old dog such as heart problems and renal failure. However, there is little information to support the view that substantial reductions in requirements are needed in healthy individuals.

Finally, there are hardly any studies at all on the vitamin requirements of elderly dogs. The particular importance of vitamin E has been cited in the scientific literature on various occasions. This vitamin is involved in detoxification processes which are believed to be increased in elderly animals. However, no studies have yet demonstrated any particular benefits of an increased intake of this nutrient in the dog.

Senior Cats

As with dog nutrition, there has been little research into the needs of elderly cats. The incidence of disease is likely to increase in the older animal and it is important that we recognise these diseases and manage them appropriately. The aim of feeding the elderly but otherwise healthy cat is to increase longevity and preserve the quality of life.

There are few obvious changes in the appearance and behaviour of the cat as it reaches the geriatric life stage but certainly there will be the same sort of physical changes in a cat as there would be in a human over the last third of its life. Indeed, there is much discussion over the subject of when cats reach 'old age'. Veterinary surgeons in general consider cats over six years of age as candidates for diseases associated with ageing (Mooney, 1991). Certainly there is fairly good agreement that when the cat reaches six to seven years of age there is a general slowing of its activity level and an increasing propensity to lay down surplus fat tissue and generally gain body weight.

As is the case for dogs, the maintenance of a steady body weight and the avoidance of obesity in the cat is certainly a critical area, as it is well accepted that obesity itself predisposes the animal to a wide range of diseases. Obesity is thought to be the most common nutritional disease in both the dog and cat (although not necessarily of particularly high incidence in the geriatric patient as previously discussed). Again there are two approaches to take in feeding, the offering of a high or low energy food. Again, the latter is probably not the correct approach as the geriatric cat may also become more fussy in its eating habits and a small amount of a highly palatable food is the preferred option. Moving from an *ad libitum* scheme of feeding to providing small but regular feeds would also be desirable so that the food intake of the cat can be carefully monitored. Hyperthyroidism has become one of the most commonly diagnosed diseases of middle-aged and elderly cats (Peterson and Graves, 1992) and is often associated with poor appetite and weight loss. Such cats can again be most effectively supported with a high energy, palatable food. Any lengthy periods of inappetence in the cat should be avoided (especially if the cat is obese) as this may cause hepatic lipidosis.

Dental calculus (tartar) and gum disease are two of the most common conditions in the older cat. These diseases often result in the loss of teeth and this should be avoided by maintaining the oral hygiene of the cat throughout its life. This may be aided by the offering of dry foods which have a naturally abrasive action on the teeth. If the geriatric cat has poor dentition it may be necessary to offer foods which are finely chopped or are further moistened. It is critical that all cats are offered a ready supply of fresh clean water and this is especially true of the elderly cat who has more problems in controlling its thermoregulation and has decreased sensitiv-ity to thirst which together may cause dehydration (Markham and Hodgkins, 1989).

As with dogs, there is little information available on changes within the digestive tract regarding efficiency of absorption, enzyme activity etc. Disorders of the digestive tract are not a serious cause for concern according to Griffiths (1968), who found that in this category of disease a considerable proportion of cases were caused by constipation and colonic impaction. It is generally thought that there will be only minor changes in the digestive tract with age but it would seem appropriate that the older animal receives foods which are of high overall digestibility. Also if there are any aberrations in digestive function it would be prudent to ensure that the intake of nutrients such as vitamins is at the upper end of the recommended range. Additionally, if the cat is on a restrictive diet or has a poor appetite, providing extra vitamins (assuming levels are within the maximum recommendations) will ensure that only energy intake is restricted and not nutrient intake. Of the diseases that occur in clinical cases the most prevalent appears to be neoplasia and there is some anecdotal evidence that vitamins A and E may inhibit certain degenerative processes.

It is currently unclear as to what are the predisposing factors in the condition of chronic renal failure in the cat. This is a fairly common disorder of the geriatric cat and there appears to be no sex or breed disposition. It has been proposed that high intakes of protein in the diet either earlier or later in life may be detrimental but there are currently no data to support this view. Certainly, there is little doubt that the treatment of this disease includes the restriction of dietary protein but it would be unwise to make the extrapolation to the healthy geriatric where restrictions might cause protein malnutrition. Undoubtedly it would be prudent to offer a food with a reasonable level of protein of a high biological value. Also implicated in renal failure are sodium and phosphorus and for these the recommendations would be similar to that of protein i.e. supply sufficient in the diet to meet the requirements of a cat which may be ingesting less food than the younger, healthy animal.

Working Dogs

Working dogs perform many different functions, for example, from acting as guide dogs for the blind to pulling sledges in polar regions. Depending on their function, working dogs have very different training, working and resting schedules necessitating different diets and feeding regimes. The amount of extra energy required depends on the environment, the amount of exercise and the nature of the work. For example, it has been calculated that a dog running for 5 kilometres will increase its daily energy expenditure by approximately 10%. Thus, a working dog travelling long distances could need as much as 2 to 3 times the normal adult ration.

Hard work induces stress, so feeding for hard work should include consideration of nutritional requirement for stress as well as fuels for muscular exercise. Many studies in humans have shown that muscle glycogen content is an important determinant of stamina and 'carbohydrate loading' is practised by many athletes to maximize muscle glycogen before exercise. For working dogs two distinct groups can be distinguished. The first one is represented mainly by the Greyhound. The effort produced by this animal during racing is very intensive but of short duration. Its muscle fibres contract very quickly and rely on glucose rather than fat to keep them functioning. A diet providing a relatively large quantity of carbohydrate would then seem suitable for this working dog. Other working dogs are active for long periods and sometimes in hostile environments (e.g. sledge dogs, avalanche dogs, etc.). They will therefore require more energy not only as a result of exercise, but also to maintain their body temperature. In all these dogs, the main fuel is fat and the muscle fibres produce energy through aerobic fatty acid oxidation. It has even been shown by Kronfeld *et al.* (1977) that a carbohydrate-free, high-fat diet appeared to confer advantages for the prolonged strenuous running performed by a group of sledge dogs.

Some studies have suggested that dogs doing very hard work may have an increased requirement for dietary protein. However,

IDEAL DIET FOR HARD WORK AND STRESS	
Energy proportions	
Protein %	32
Fat %	51
Carbohydrate %	17
Dry matter basis	
Protein %	42
Fat %	30
Carbohydrate %	22
Fibre%	2
Ash %	4
Digestibility %	90
Main ingredients	Meat, meat by-products, grain

TABLE 5.6: Ideal diet for hard work and stress.

there is no scientific evidence to suggest that feeding a high protein diet will promote superior muscle development. Table 5.6 shows the composition of an ideal diet for hard work and stress as suggested by Kronfeld (1982).

There is very little information available about the requirements of hard-working dogs for minerals and vitamins. If the diet is nutritionally balanced, the dog will eat to satisfy its energy requirement and will simultaneously consume all the minerals and vitamins it requires. There may be a higher iron requirement in the hard-working dog because of its involvement in haemoglobin production and oxygen transport but this has never been studied in detail, as typical working dog diets are high in meat and will therefore be high in iron. It has also been postulated that the working dog may require more vitamin E and selenium than the normal adult to prevent red blood cell fragility but this too has not yet been studied in detail.

The feeding regime of working dogs is as important as the diet itself. All working dogs should receive only a small meal before working, since a full stomach is not conducive to efficient work. In addition it is often inconvenient for guard and guide dogs to have to defaecate during a duty period. The main meal should provide two thirds of the daily needs and should be given after a rest period. Working dogs should be given ample opportunity to drink during the working period.

In summary, foods for working dogs should be highly palatable, concentrated, digestible and nutritionally balanced. Such foods do not need further supplementation.

CHAPTER 6

Nutrition of Pet Birds

HELEN M. R. NOTT and E. JEAN TAYLOR

Introduction

The ownership of pet birds varies markedly throughout the world, both in the numbers of birds in total and the variety of species kept. The budgerigar (*Melopsittacus undulatus*) and the canary (*Serinus canarius*) are undoubtably the most popular species. The budgerigar was first brought to Britain from Australia in the 1840s and, following successful breeding, was rapidly introduced to other European countries. The canary has a much longer history as a cage bird and may well have been kept by the inhabitants of the Canary Islands prior to the Spanish conquest in the 15th century (Zeuner, 1963). The Spanish in the 17th century and the Italians in the 18th century dominated canary breeding where the males in particular were valued for their song (Ellis, 1984). The budgerigar is more popular in Northern European countries (about 65% budgerigars versus 15–20% canaries of the total number of birds kept) whereas the canary predominates in southern European countries such as Spain and Italy. The remainder of pet birds consist of a variety of exotic finches, parrots and parrakeets and a much smaller number of others species such as pigeons and quails. In Australia and the USA the species kept are again very similar but budgerigars tend to be less popular in Australia, where they are native and the large parrots are more popular in both countries than in Europe.

Feeding Behaviour

Feeding Ecology

The majority of commonly kept companion birds are natural seed-eaters although some have more specific diets; such as the lories and lorikeets which are nectar feeders.

Canaries were originally derived from wild species found in the Canary Islands, Madeira and the Azores. They are primarily seed eaters, however, no scientific studies on their feeding ecology in the wild have been conducted to confirm their natural diets. Of the exotic finches the most commonly kept is the zebra finch (Ellis, 1984) and, typical of most finches, they are seed eaters derived from a wild grassfinch in Australia. Zann and Straw (1984) observed the feeding ecology of wild zebra finches in farmland. The major limitation on colony numbers appeared to be the presence of suitable bushes and trees for nesting and roosting, rather than food supply. Examination of crop contents revealed a wide variety of seed types, both wild and agricul-

NUMBER OF PLANT SPECIES ON WHICH WILD PSITTACINES HAVE BEEN OBSERVED FEEDING		
	No. of species	**Reference**
Budgerigar (*Melopsittacus undulatus*)	21–39	Wyndham, 1980a
Red-capped Parrot (*Purpureicephalus spurius*)	28	Long, 1984
Port Lincoln Parrot (*Barnardius zonarius*)	>41	Long, 1984
Western Rosella (*Platycercus ictereotis*)	43	Long, 1984
Eastern Rosella (*Platycercus eximius*)	82	Canon, 1981
Pale-headed Rosella (*Platycercus adscitus*)	47	Canon, 1981
White-tailed Black Cockatoo (*Calyptorhynchus funereus*)	30	Saunders, 1980

TABLE 6.1: The number of plant species on which wild psittacines have been observed feeding. (Data taken from Nott, 1992.)

tural. Insects and snails were taken by only 0.07% of the birds observed. Presumably by taking a wide variety of seeds the birds are able to adapt to changing availability and can meet their nutritional needs.

The budgerigar is one of the smaller parakeets native of Australia. Wyndham (1980a,b) studied the feeding behaviour of wild budgerigars in inland mid-eastern Australia. Large flocks were observed feeding in the early morning and late evening, with smaller flocks forming in the middle of the day. They fed on 21–39 species of ground plants either directly from the seed heads of grass tussocks or from seeds that had been scattered on the ground. Studies of other parakeets and parrots in Australia have shown a similar, or higher, diversity of seed types taken (Table 6.1). There have been relatively few studies on the feeding ecology of parrots from Africa or South America, probably because the greater density of foliage in their habitats makes observations more difficult. Anecdotal data suggest that these species take a wide variety of food types including seeds, fruit, nuts and shoots and studies on the Bahama amazon (*Amazona leucophala baharensis*) have shown at least 16 different plant species ingested (Snyder *et al.*, 1982). In summary, the psittacines (parrots and parakeets) seem to be very catholic in their food choices, presumably adapting to changing food supplies in the wild as different species of plants come in and out of season.

By taking a wide variety of foods their nutritional needs are also more likely to be met than if they had a more conservative diet.

Food Preferences and Feeding Patterns

In addition to field studies a number of investigations have been conducted into the feeding preferences of birds in captivity. Studies on the seed preferences of a wide variety of finches have found that the species with larger beaks, rather than larger bodies, preferred the larger seeds (Morris, 1961; Kear, 1962). Species or individuals with larger beaks presumably find it easier to handle larger seeds and will therefore be able to dehusk and consume the kernel more quickly. It is therefore more profitable, in energy terms, to ingest larger seeds than a greater number of smaller ones. By experience each bird will learn to take seeds which are the most profitable and this will explain the preferences observed by Morris (1961) and Kear (1962). Rabinowitch (1969) observed that experience, particularly early in life, could influence a finch's seed preferences. Birds preferred their rearing diet, especially in the first five weeks of fledging. After this time their preferences were modified, presumably by learning, from other seeds sampled. Our own studies have shown similar effects in budgerigars.

Birds which have been raised on diets of minimal variety may be less willing to accept novel foods (neophobia) than those with a greater experience of different food types, a factor we have observed in our own studies on budgerigars. This can lead to problems in adulthood when birds become fixed on a limited range of foods and nutritional deficiencies develop. Mixing new food items in with familiar food is the most effective way of ensuring acceptance although birds can still be very adept at selecting out preferred items. It is therefore prudent to provide all birds with a wide variety of foods immediately after fledging to minimise the risk of neophobia developing.

The feeding patterns of pet birds have been relatively poorly studied. Slater (1974) studied the feeding patterns of zebra finches in captivity and recorded significant individual variation. Peaks in feeding occurred in the first hour after the lights came on and just before the lights went out. This matches observations on wild zebra finches which tended to feed primarily after dawn and just prior to roosting (Zann and Straw, 1984), presumably replenishing depleted reserves in the morning and 'topping' up reserves in anticipation of a night without foraging. Our own studies show similar patterns in pet budgerigars, although the feeding patterns tended to be less consistent than those of the zebra finches and meal sizes were more irregular. This may be a reflection of the smaller body size of the finches and the need to regulate energy balance more closely.

Many larger species of psittacine are commonly fed distinct meals. This method minimises the risk of obesity, since the total amount offered can be controlled and also encourages the birds to become less selective in the specific food items that are taken from a seed mixture. An additional advantage is that the routine can help to establish a closer relationship between the bird and its owner since the bird will look forward to each interaction in anticipation of a food reward.

The Avian Digestive System

In comparison with most other animals birds have a particularly high metabolic rate. Thus they have evolved a digestive system which is capable of digesting food quickly and efficiently and which differs in many respects from the digestive systems of other species.

To enable birds to break up large pieces of food or remove the outer husks from seeds the avian jaw has developed into a beak which is a hard, keratinised structure. In the case of the budgerigar the lower, horseshoe-shaped beak fits into the upper, heavily keratinised beak, an ideal structure for dehusking seeds. The dehusking is aided by palatal pads on the roof of the mouth which allow the husk to be cracked, removed and ejected while the kernel is retained and swallowed (Evans 1969). The beak grows constantly and must be worn by normal use in order to maintain optimal function. A second difference is the absence of teeth in the avian jaw. This has two advantages for the bird: firstly it eliminates the action of chewing food which means that ingestion is a much more rapid process. This allows more time to be spent watching other birds in the flock or potential predators. Secondly, the lack of teeth is a weight-saving adaptation which has obvious benefits in terms of flight. The absence of the musculature required for chewing also contributes to a reduced body weight (Brooke and Birkhead, 1991).

In most species, including psittacines, once food is swallowed it passes into a storage organ known as the crop. This is another evolutionary adaptation to permit rapid feeding. The crop is a reservoir for storing and soaking food and plays a minor role in enzymatic digestion and absorption. There is, however, evidence that some microbiological activity takes place within the crop which results in the production of organic acids. If the bird's stomach is empty, then food bypasses the crop and travels directly to the proventriculus (see below). Certain species, primarily pigeons and doves, possess crops which are adapted to produce 'crop milk'. This is not milk as such but is, in fact, the epithelial lining of the crop which is sloughed off and fed to the young birds for the first three days of their lives. On a dry matter basis crop milk contains almost 60% protein with about 35% fat and 5% minerals but unlike mammalian milk has no casein or lactose (Hegde, 1973).

Following storage in the crop, peristaltic action forces food into the stomach. If the stomach is full then peristaltic action ceases for 30–40 minutes to allow emptying (Sturkie, 1954). In birds the stomach comprises two chambers—the proventriculus and the ventriculus. In the proventriculus there is an abundance of chief cells which secrete both pepsinogen and hydrochloric acid. This initiates the process of digestion which continues as the food passes into the ventriculus. In some bird species, including budgerigars, newly hatched birds are fed on a secretion from the hen's proventriculus (rather than the crop) which is often incorrectly called crop milk. This fluid, like crop milk, has a particularly high protein content, essential for the young, rapidly developing bird.

The ventriculus, commonly called the gizzard, functions to break down food by both chemical and physical means. The latter is achieved by powerful muscle contractions and the presence of insoluble grit. Birds ingest grit (see later) for this purpose and have a thick gizzard lining to prevent damage to the tissue. This is not composed of keratin, as has frequently been suggested, but appears to be a polysaccharide–protein complex known as koilin (Petrak, 1982). This 'koilin layer' is particularly well developed in birds which feed on hard foods such as seeds or cereals. The presence of sand or grit in the gizzard increases the motility and grinding action by forming an abrasive background for the muscular action. Grit may be required for optimum digestion in the bird and studies in poultry have shown it to improve the digestion of whole seeds and grains by as much as 10% (Sturkie, 1954).

Thus it could be said that the three organs—the crop, proventriculus and gizzard—combine the equivalent functions of the teeth and stomach of monogastric animals. The subsequent stage of digestion is similar to the mammalian process. Digesta pass into the small intestine where pancreatic juice and bile are added at the distal end of the duodenum. This completes the breakdown of most food constituents which are then absorbed across the wall of the intestine. In some species, such as fowl, there is further degradation by microbial fermentation in the paired caeca which are relatively large. However, many species possess either very small caeca such as the canary in which they are vestigial, or, as in the case of the budgerigar, are completely absent. Thus for these species, digestion is complete by the time food reaches the large intestine. In fact it appears that the only function of the large intestine is water reabsorption. The large intestine is relatively short and only slightly wider than the distal small intestine. It empties directly into the cloaca from which the faeces are excreted. The gut transit time of food may be anything between 1.5 and 12 hr depending on the bird species, the type and amount of food and the physiological state of the bird (Sturkie, 1954). One of the fastest food transit times is that of the toucan which is reported to have a total transit time of 1.3 hr (Honigmann, 1936). For companion birds such as budgerigars and finches transit times are longer—between 3 and 6 hr.

In terms of the excretory processes there are two further differences between mammals and birds. In mammals arginine is used for the conversion of ammonia to urea. This occurs in the liver via the ornithine–arginine cycle, the first stage of which is catalysed by the enzyme arginase. Birds' livers do not contain arginase, therefore excess nitrogen is excreted as uric acid which is formed in the liver and kidneys. Whereas in mammals up to 90% of urinary nitrogen is urea, in birds between 60 and 80 % is uric acid. The production of uric acid is a more complicated process and is more expensive in terms of energy utilisation (Buttery and Boorman, 1976). However, the excretion of an insoluble nitrogen compound means that birds have a low requirement for water. Again this is advantageous in terms of maintaining a low body weight. Urine and faeces are voided together via the cloaca where the ureters and large intestine both exit.

Nutritional Requirements Of Companion Birds

Since few nutritional studies have been carried out with parrots and parakeets (Psittaciformes) or finches and canaries (Passeriformes), in order to understand their nutritional requirements it is necessary to extrapolate from the available data on poultry. These species belong to the orders Anseriformes and Galliformes and as such are biologically distinct from both the Psittaciformes and the Passeriformes. Applying the results of studies conducted with poultry must therefore be done with caution. Additionally, poultry are fed in such a way as to achieve maximal growth rates and optimal carcass composition. These objectives are somewhat different from those of pet bird owners.

Studies on passerines, such as wild finches and sparrows, often provide nutritional information for canaries and finches. Furthermore, knowledge of the nutritional requirements of pigeons may contribute to our understanding of the nutritional needs of other companion birds. It is for these reasons that the nutritional recommendations quoted in this chapter often refer to birds other than those which are typically kept as pets.

Energy

Energy Requirements for Maintenance

The primary energy source in the diet of seed-eating birds is carbohydrate, mainly in the form of starch. Fat also provides a highly concentrated energy source and can be particularly important for newly-hatched chicks whose energy demands are high. Pigeon squabs (chicks), for example, satisfy their energy requirements with the fat found in crop milk (Griminger, 1983). If there is insufficient dietary carbohydrate or fat then protein can also be used as an energy source. Like other species, birds eat primarily to satisfy their energy requirements. Thus the energy density of the diet is an important consideration since birds adjust food intake to provide a constant energy intake. Studies with poultry have shown that if the energy content of the diet is increased, for example by increasing the proportion of fat, then birds decrease intake to compensate for this. If no changes are made to the other nutrients then eventually deficiencies develop. Conversely, where

the dietary energy density is diluted, then intake of the diet increases so that the bird meets its energy requirement for maintenance. However, if the diet is excessively dilute, the ability to adjust intake may be overridden because gastrointestinal capacity becomes a limiting factor (NRC, 1984).

The daily metabolisable energy (ME) requirement of an adult domestic fowl has been calculated as 300 kJ/kg body weight (McDonald *et al.*, 1988). When this is adjusted for metabolic body weight the figure becomes 360 kJ/kg $W^{0.75}$ (where W = body weight). If this value is applied to the budgerigar, assuming an average bodyweight of 53 g, the ME requirement equates to 40 kJ/d. The validity of extrapolating poultry data to other avian species is questionable and it is probably more appropriate to apply results obtained from studies on passerines.

Specific work on finches (Aschoff and Pohl, 1970) showed their ME requirement to be $535 W^{0.715}$ kJ/kg/day. For a budgerigar weighing 53 g this equates to 65 kJ per day. Studies on the energy intake of adult budgerigars (Earle and Clarke, 1991) measured the daily ME requirement as 90 kJ for a 53 g bird. Clearly there is a discrepancy of about 25% between these two values but when the conditions of the studies are taken into consideration the reason for this becomes clear. Aschoff and Pohl measured the maintenance energy requirement of birds which had been fasted prior to the study whereas Earle and Clarke studied budgerigars under normal conditions. When food is eaten the body utilises between 10 and 15% of ME in order to assimilate the food—the so-called thermogenic effect of food (Calles-Escandon and Horton, 1992). Considering the frequency with which birds feed—budgerigars have approximately 14 meals per day—there is a continuous energy input into this process. Furthermore, birds expend a relatively high proportion of ME in the formation of uric acid (the end-product of protein breakdown). This equates to approximately 1400 kJ/mole of uric acid synthesised: a highly energy intensive process compared with that of urea synthesis which utilises 430 kJ/mole of urea formed (Buttery

METABOLISABLE ENERGY REQUIREMENTS OF SOME COMPANION BIRDS		
Bird	**Weight range**	**Daily ME req (kJ/d)**
Cockatiel	80–100g	110–130
Budgerigar	50–70g	78–100
Canary	20–30g	40–55
Zebra finch	15–20g	32–40

TABLE 6.2: Metabolisable energy requirements.

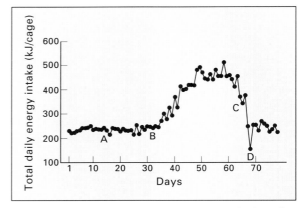

FIG 6.1: Mean daily energy intake (kJ/cage) for breeding pairs of budgerigars throughout the 78 day breeding period. A is point at which first egg is laid, B is point at which first chick hatches, C is point at which first chick fledges, D is point at which last chick fledges.

and Boorman, 1976). Thus ingesting and digesting food, as well as excreting the breakdown products probably increases the ME requirement by 25%. Applying the Aschoff and Pohl equation to other species and allowing a 25% increment for food assimilation gives us a good idea of the daily energy requirements of companion birds (Table 6.2). In terms of $W^{0.75}$ (kg) the daily energy requirement constant (kJ) is around 740, approximately double the value reported for the domestic fowl.

Since the energy content of most budgerigar seed mixes is around 1.75 MJ per 100 g, theoretically a bird would need to eat roughly 4 to 6 g per day in order to meet its energy requirements. However, as the bird discards the husk of the seed then enough whole seed must be offered to provide 4–6 g of kernel. Furthermore, since the energy digestibility of seeds is normally around 90% then this increases the quantity of seed required to provide sufficient energy. Studies have shown that on average a budgerigar consumes between 8 and 12 g of seed per day (Earle and Clarke, 1991) which is equivalent to between 500 and 1000 individual seeds.

Energy Requirements During Breeding

The average weight of a budgerigar egg is 2.5 g with every gram containing around 6.5 kJ of energy. Therefore one budgerigar egg contains about 16.5 kJ. With an average of 5 eggs per clutch the hen may require over 80 kJ of extra energy during the laying period. Furthermore, since energy is required by the hen to produce the eggs then this value is even higher. Clearly, more food must be offered to compensate for this increased demand. Measurements of energy intakes by

a pair of breeding budgerigars over the rearing period (78 days) are shown in Fig. 6.1 (Earle and Clarke, 1991). During the period of laying and incubation, energy intakes remained relatively constant at 231–252 kJ per day. However, once the chicks began to hatch the energy requirements of the parent birds increased dramatically. At peak intake the birds collectively ingested between 483 and 505 kJ per day (2 adults with 3 chicks). As well as the extra food required for the rapidly growing chicks, the adult birds expend extra energy actually carrying food and feeding it to the young. In these studies this resulted in an average weight loss of 8 g by the cock birds.

Energy Requirements for Growth

At hatching a budgerigar chick weighs about 1.5 g yet by the time it is 10 days old its weight is around 20 g. In order to achieve this phenomenal growth rate, as well as develop a complete covering of feathers, the young bird has a very high energy demand. Young birds devote a large proportion of their energy intake to supporting growth. This has been measured as between 62 and 73% in robin nestlings aged 1–3 days. This proportion gradually decreases as the nestling develops, but over the entire period from hatching to independence an average of 11% of the daily energy budget is used for growth (Brooke and Birkhead, 1991). Consequently extra food should be supplied to the adult birds so that they can supply the chicks with their nutrient requirements. On average the food requirement will double during the rearing period and will only return to pre-hatching levels once the last fledgling has left the nest.

Effect of Temperature

Birds maintain their body temperatures at around 41–42°C—normally much higher than environmental conditions—and a high energy input is required to achieve this. The energy required to maintain a constant body temperature depends upon the difference between the normal body temperature and the environmental temperature. Since companion birds tend to be small, the surface area of the body is large in relation to body weight. This means that in a cold environment the bird must employ an efficient mechanism in order to conserve heat and food intake will be at a maximum. Equally, when the surrounding temperature is particularly high, heat must be disposed of. Birds reduce their body temperatures via the evaporative routes of the respiratory tract and skin. However, they are relatively inefficient at this process and in a hot environment often decrease heat production by reducing food intake. Food intake decreases by about 1.5% for each rise of 1°C above the normal temperature range (NRC, 1984).

Fat and Essential Fatty Acids

As with other animal species fat has two major roles in avian nutrition: it provides a concentrated energy source and performs various metabolic functions as well as assisting in the absorption of the fat-soluble vitamins. The metabolism of fat to provide energy occurs by the same pathway as with other animal species: breakdown by lipases during the process of digestion produces glycerol and fatty acids. These then undergo a series of reactions to provide energy stored in the form of adenosine triphosphate. Birds are sensitive to the dietary level of fat and will regulate their metabolism to prevent excessive energy storage. This is achieved by maintaining a constant feed intake in order to satisfy other nutritional requirements but metabolising the fat less efficiently. The overall effect, therefore, is weight maintenance and not weight gain. However, birds fed on high-fat diets will become obese since they must continue eating to meet other nutritional requirements. Furthermore, a high fat intake can cause diarrhoea and may also result in the formation of insoluble soaps rendering minerals such as calcium and iron unavailable to the bird (Wallach, 1970). Some cockatoos fed solely on sunflower seeds become obese and exhibit a high incidence of lipomas. This is frequently seen in rose-breasted cockatoos and underlines the importance of providing a diet based on a mixture of seeds (Harrison and Harrison, 1986).

In terms of their metabolic roles fatty acids perform a variety of functions. 'Parent' essential fatty acids (EFA) may undergo desaturation and chain elongation leading to the production of long chain polyunsaturated fatty acids (PUFA). PUFA form an integral part of membranes and lipid transport systems while the EFA are necessary for prostaglandin synthesis. Generally linoleic acid is considered to be the quintessential fatty acid for birds since it will eliminate all signs of EFA deficiency (Watkins, 1991). The first sign of EFA deficiency in chicks is retarded growth which can manifest itself within one week of feeding a deficient diet. This is followed by the deterioration of membrane structures which results in dermal problems. The skin becomes rough, flaky and increasingly permeable resulting in rapid water loss. This forces the bird to increase its water intake—a classic sign of EFA deficiency. If EFA are not provided then feathering deteriorates, feed utilisation is impaired, resistance to disease falls and eventually the bird dies. In adult birds this is extremely uncommon since the body has a high capacity to maintain reserves of EFA. Although it appears that young chicks cannot absorb fat as well as adult birds, in general both immature and mature birds have similar EFA requirements. Studies on poultry suggest that about 2% of dietary ME should be provided as linoleic acid. This corresponds to about 0.9% of a diet containing 12.5 MJ/kg. It is possible that there is a significant increase in the need for dietary fats during periods of moulting in adult birds. The membranes of the epithelial cells used in feather production have a significant fat component which must be supplied in the diet. Good sources of linoleic acid include oil seeds such as linseed.

Protein and Amino Acids

Protein Requirements for Adult Birds

Like all animal species birds require protein for the formation and development of body tissues and the maintenance of body structure. In addition, birds utilise protein for the production and maintenance of feathers, claws and beaks. Poultry studies have established that a dietary protein content of between 15 and 18% (as is) is optimal for adult birds (McDonald *et al.*, 1988) while studies with adult pigeons indicate that a dietary content of 12.5–13% is adequate (Wolter *et al.*, 1970). Further studies with budgerigars have established that there is a minimum dietary requirement of 10% protein (Drepper *et al.*, 1988). Work with passerines (adult tree sparrows) confirmed that a dietary protein content of 8–9% is sufficient to maintain weight, nitrogen and energy balance (Zazula, 1984). However, this figure is probably below the requirement for optimal growth, development and repair. This is particularly true during certain life stages such as chick development, or in the case of adult birds, during moulting. Hens also have a high protein demand during the reproductive season for the production of eggs and, subsequently, 'crop milk'. Studies with pigeons have shown that breeding birds require a dietary level of 18% protein (Wolter *et al.*, 1970).

Additionally, the dietary protein requirement is dependent on food intake. Thus, anything which affects food intake such as the energy content of the diet or environmental temperature, indirectly affects the optimum dietary level. There is no advantage in feeding high quantities of protein to birds since the energy costs of deamination, uric acid formation and excretion are very high. Furthermore, an overload of the excretory ability of the kidneys may lead to hyperuricaemia. The subsequent deposition of urate crystals in the kidneys can produce clinical symptoms and may cause death (Lowenstine, 1986). There is also a possible link between excessive dietary protein and overgrowth of beaks and claws in cage birds (Hasholt and Petrak, 1983).

Protein requirements have both a quantitative and a qualitative aspect. In terms of the former, there must be enough dietary protein to supply the non-essential amino acids or sufficient nitrogen for their synthesis. From a qualitative point of view, dietary protein must provide the amino acids which the bird is unable to synthesise or cannot synthesise rapidly enough—the essential amino acids. In addition to the essential amino acids required by mammals, it appears that chicks may need a dietary source of glycine and proline in order to achieve maximum growth and development (see below).

The maintenance of an optimum balance between amino acids is extremely important for the efficient utilisation of dietary protein. Ideally the diet should meet the requirements for all the amino acids without including either excess protein or individual amino acids. A low protein diet, causing a moderate deficit in all the amino acids results in increased food intake. This is an attempt by the bird to compensate for the inadequate nature of the diet but ultimately results in obesity. An even more marked response is seen if just one or two amino acids are fed in abnormally low or high concentrations. Under normal circumstances deficiencies in individual amino acids are rarely seen. However feeding cage birds on a narrow range of seeds or grains may induce deficiencies of certain amino acids, particularly lysine and methionine which tend to be low in certain seeds, for example, safflower seeds and peanuts (Lowenstine, 1986).

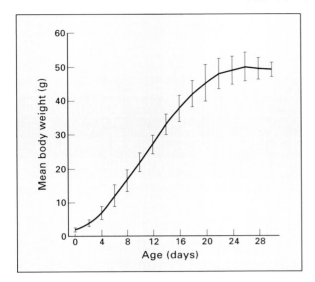

FIG 6.2: Body weight increase (g) for budgerigar chicks ($n = 16$) from hatching to day 30. Values are means ± SD.

Protein Requirements for Growth

The demands of rapid growth and development mean that young birds have high protein requirements. It is hardly surprising that the chick's protein requirement is so much higher than that of the adult bird when its rate of growth is taken into account. As shown in Fig. 6.2 the newly-hatched budgerigar chick doubles its weight within two days and then continues to grow rapidly until it is about thirty days old. In addition to the huge weight gain the bird also develops a complete feather covering. However, the optimum dietary protein level has yet to be established. It appears that when the protein level is at least 25%, chick development is noticeably better. The liquid produced in the budgerigar crop for the chick contains between 24 and 26% protein as a proportion of dry matter which would suggest that this is an optimum level for newly-hatched budgerigars. After a few days the requirement appears to decrease slightly and studies with pigeon squabs have determined that, at this stage, optimal growth and development is achieved when the dietary protein content is 18% (Wolter *et al.*, 1970). Studies with cockatiel chicks have established that the diet should provide 20% protein in order to support maximal growth (Roudybush and Grau, 1986) and further studies suggested a minimum requirement of 0.8% lysine (Roudybush and Grau, 1991). Recent studies with growing canaries suggest that the dietary protein content should lie between 16.5 and 21.9% (Kamphues and Meyer, 1991). Excessive dietary protein (23% of diet) has been shown to depress growth in older cockatiel chicks so it is wise not to over-supplement diets (Roudybush and Grau, 1986). Many bird owners offer breadcrumb mixes as a soft food for breeding birds. However, since the protein content of wheat is, like most seeds, around 14% this has no advantage as a high protein source. An alternative is to offer bread soaked in milk. Although this has a higher protein value the amino acid balance is not ideally suited to young birds. Therefore it is wisest to offer a diet which is specifically formulated for growing birds and has a protein content of around 20%. Such diets are usually egg based and can therefore provide both an optimum protein supply and the correct balance of amino acids. As the chick develops, its requirement decreases and its diet alters accordingly. After a few weeks, when the bird has achieved its adult weight (about six weeks for the budgerigar) its protein intake should be the same as a normal adult bird.

The amino acid requirements of developing chicks appear to be different from those of the adult bird. As well as the 10 essential amino acids needed by all birds, poultry chicks require a dietary source of both glycine and proline for optimum development (Featherstone, 1976). Glycine is an integral precursor of uric acid—one of the reasons why birds have a high dietary requirement for this amino acid. Additionally, glycine is a major component of collagen and feathers (Table 6.3). Thus a deficiency of this particular amino acid is quickly apparent in the developing chick. If the glycine demand exceeds dietary supply, serine can be converted to glycine in the liver. However, since the rate at which the chick synthesises glycine is slower than the rate at which it is used then, to ensure optimal growth, a dietary supply is

| MAJOR AMINO ACIDS IN BUDGERIGAR FEATHER PROTEIN ||
Amino acid	% of Protein
Glutamic acid	7.72
Cystine	7.45
Proline	7.26
Valine	5.41
Leucine	5.30
Glycine	4.60
Aspartic acid	4.54
Arginine	4.27
Serine	4.12
Alanine	4.07

TABLE 6.3: The major amino acids of budgerigar feather protein.

beneficial (Snyder and Terry, 1986). A similar situation exists for proline. This is a major constituent of both collagen and feathers which accounts for its essential nature in the chick's development. Glutamate is the usual precursor of proline and, for this reason, is often quoted as being 'semi-essential' in the diet of the chick.

Protein and Amino Acid Requirements During Moult

Plumage provides birds with thermal insulation as well as acting as a water repellant and an aid to flying. Formation of the follicles occurs during embryo development but, subsequently, feathers are shed and regenerated throughout the bird's life. At four months the budgerigar goes through its first moult. From then on, in Western Europe, the main annual moult is in the autumn. This begins when the temperature falls and usually continues for six to eight weeks. The pattern is the same for canaries and finches.

Feathers comprise between 85 and 97% protein—almost all of which is keratin. Thus dietary amino acids play a critical role in feather development. The major ones involved in feather keratin synthesis are the sulphur amino acids—methionine and cystine. Cystine is a primary component of keratin (Table 6.3) while methionine is important as a cystine precursor. Studies have shown that during periods of rapid feather development, maximum feed efficiency is obtained if at least

half of the dietary sulphur amino acids are present as cystine. Once feather growth is complete the dietary value of cystine decreases (Deschutter and Leeson, 1986). During periods of moult, therefore, it is advisable to offer seeds such as rapeseed and white millet which contain high levels of these particular amino acids (Table 6.4). There are other amino acids which are vital for normal plumage including lysine and arginine. It appears that lysine is required for pigmentation, possibly due to its role in melanin formation. There is also evidence of a link between lack of dietary arginine and feather plucking (Buttner, 1968).

Birds which lose feathers, either as a result of disease or plucking, have higher nutritional demands to compensate for this. In addition to the amino acid requirements for new feather growth, partially feathered birds experience increased heat loss. This results in an increase in the metabolic rate of up to 60% with a concomitant increase in energy demands for heat production and feather regrowth. Energy intake by defeathered birds is about 85% higher than for those with normal feathering (at 22°C) (O'Neill et al., 1971). Thus at such times it might be advantageous to feed seeds with higher energy densities such as groats or oilseeds.

Amino Acid Interactions

The glycine–serine and glutamate–proline inter-relationships demonstrate the importance of considering amino acid requirements within the context of the whole diet. For adult birds one of the most significant inter-relationships is between the sulphur amino acids which are integral components of feathers (see above). In view of the interactions between methionine and cystine, therefore, the minimum dietary requirement for methionine is expressed as a requirement for both these amino acids. Similarly, for chicks the dietary requirement for glycine is stipulated in terms of the total glycine and serine content of the diet. To complicate the situation further, the glycine requirement is increased by low dietary concentrations of methionine, arginine

THE AMINO ACID COMPOSITION OF DIFFERENT SEED TYPES						
Amino acid	Canary seed	White millet	Groats	Niger	Red millet	Rape
Cystine	3.46	1.34	3.44	2.42	1.71	2.91
Methionine	1.90	5.79	1.72	2.37	3.06	2.16
Threonine	2.20	3.05	3.36	3.17	5.94	4.47
Serine	3.77	6.17	4.92	4.80	2.79	3.92
Proline	6.68	6.26	4.59	3.48	6.93	7.23
Glycine	2.62	2.15	4.59	4.53	2.34	4.57
Valine	3.69	3.90	4.26	3.52	3.24	4.12
Isoleucine	3.53	3.38	3.27	3.43	2.43	3.06
Leucine	6.59	11.77	6.47	5.94	11.53	6.08
Tyrosine	3.01	3.56	3.52	2.73	3.33	2.46
Phenylalanine	17.26	17.98	14.67	15.06	5.31	16.43
Lysine	2.63	1.98	3.93	4.36	1.08	6.53
Histidine	2.09	2.40	2.37	2.33	1.89	3.26
Arginine	6.78	3.29	7.78	6.43	3.06	5.12
Total protein	14.7	11.2	12.2	22.7	11.1	19.9

TABLE 6.4: The amino acid composition of different seed types.

or B-complex vitamins (McDonald *et al.*, 1988).

Since the requirements of individual amino acids have yet to be conclusively established for companion birds the importance of feeding a mixed diet cannot be over-emphasised. As can be seen from Table 6.4, the protein and amino acid contents of the various seed types are significantly different. Niger, for example, has a good amino acid profile but an excessively high protein content with respect to birds' requirements. It would therefore be unwise to feed it solus to a bird. A similar situation exists for rapeseed which, again, has a relatively high protein content but is a particularly good source of both proline and histidine. Ideally, then, seed-eating birds should be offered a diet based on a good range of seeds.

Carbohydrates

The major function of all ingested carbohydrate is to provide a source of immediate energy to the bird. There are no significant differences between avian species and other species in the ways by which they utilise carbohydrate. The enzymes of the digestive tract are as efficient at breaking down starch as those of other animals. However, the ability to digest non-starch polysaccharides is more limited, particularly in birds whose caeca are vestigial or completely absent.

Water

Physiologically birds are less dependent than mammals on drinking water, since they eliminate nitrogenous waste as insoluble uric acid and not urea (see above). Consequently very little water is contained in the excreta—an average of 32.8 ± 6.5% (Earle and Clarke, 1991). Nevertheless water remains an essential dietary constituent. It acts as both a transport medium and a site for metabolic reactions and is vital for the regulation of body temperature. The avian species for which a water supply is most critical are those whose staple diet is dry seeds. Thus companion birds such as budgerigars, canaries and finches require a regular supply of fresh water. Budgerigars receiving a supplemented seed diet consume between 3 and 5ml of water every day (Earle and Clarke, 1991). Of cage birds, canaries appear to be the most sensitive and can die within 48 hours if deprived of water. Budgerigars, on the other hand, have evolved to survive on minimal water but should still be supplied with fresh water on a daily basis. If a bird does not have an adequate water intake then its droppings become small and scant. Additionally the proportion of the droppings that is faecal matter increases so they appear to be dark green in colour.

Grit in the Diet of Caged Birds

There are two types of grit which a bird may

ingest in its diet—insoluble, e.g. quartz, and soluble, e.g. oyster shell. The former may aid digestion while the latter may provide a source of certain minerals. Frequently the functions of these types of grit are confused so it is important to recognise the role that each has in avian nutrition.

Insoluble grit. Poultry studies have established that dietary grit may improve the digestibility of mash by around 3% and whole grains and seeds by up to 10% (Sturkie, 1954). However, it is not indispensable since chicks can be raised to maturity without dietary grit with no adverse effects on growth and development. Since grit is retained in the digestive system then it is unnecessary to provide it continuously. It has been suggested that providing psittacines with 10–12 grains 2–3 times a year is sufficient for the efficient functioning of the gizzard (Harrison and Harrison, 1986). However, the essential nature of grit in the diet of psittacines has yet to be established and increasing evidence suggests that there is no requirement for a dietary supply. Oversupplementation of grit may lead to health problems such as crop impaction. This is becoming increasingly common in psittacines and as a result some vets recommend that no grit be supplied to pet birds (Evans, 1992). If grit is to be provided in the diet then it is important to ensure that the particles are a manageable size for the bird. Obviously very large particles will be difficult for the bird to swallow whereas particles which are too fine lead to impaction of the gut.

Soluble grit. If there are insufficient minerals in the seed diet then soluble particles of grit can provide a valuable source of minerals such as calcium and phosphorus. Since soluble grit is readily degraded and absorbed in the avian digestive system, it does not assist in the digestive process.

Minerals

Calcium and Phosphorus

Calcium and phosphorus are essential for the development and maintenance of the skeletal structure. In the growing bird most of the calcium in the diet is used for bone formation, whereas in breeding most will be used by the hen in egg shell formation, which is around 98% calcium carbonate. Due partly to the rapid growth rate of species of caged birds the calcium and phosphorous requirements are high and the ideal ratio of calcium to phosphorus (Ca:P) is also particularly high (around 2:1) when compared with most mammals.

An imbalance in the absorption of calcium and phosphorus (or the level of vitamin D in the diet) is quite common in pet parrots and results in a variety of disease conditions. The most frequently seen of these is rickets. This generally arises as a result of the provision of only high fat seeds which tend to have a low mineral content and a low Ca:P ratio. In seeds with high oil contents insoluble calcium soaps can form in the intestine limiting calcium absorption and further exacerbating the absolute and relative calcium deficiencies (Wallach, 1970). To preserve normal blood calcium levels skeletal calcium is solubilised leading to osteomalacia. Chronic deficiency usually results in nutritional secondary hyperparathyroidism, a relatively common disease of parrots (Wallach and Flieg, 1967). Birds maintained on a mildly deficient diet can take years to develop clinical signs of disease which are usually in the form of loss of appetite, feather chewing, weakness and lethargy. In more marked cases bone fractures can occur and some individuals may develop hypocalcaemic tetany and seizures (Randell, 1981). There is some evidence that some species of psittacine are more susceptible to calcium deficiency than others. Murphy (in Rosskopf and Woerpel, 1991) studied African grey and timneh grey parrots, which show a high incidence of hypocalcaemia and concluded that they are unable to mobilise calcium from their skeleton and so periods of low dietary calcium result in lowered blood calcium levels.

In practice adequate calcium can be provided by the provision of shell sand, limestone grit or cuttlefish or direct supplementation of the diet. High concentrations of calcium carbonate or calcium phosphates in the diet may make it unpalatable and so care should be taken in the levels offered. Field experience with diets of 1% calcium has suggested this is sufficient for reproduction in some of the larger species of psittacine (Ullrey *et al.*, 1991) and is the level recommended by Perry (1983) to prevent calcium deficiency. In contrast, 0.35% has been found to be adequate for the reproducing cockatiel in maintaining shell thickness and egg shell conductance (Roudybush and Grau, 1991). Whilst our own studies have shown that 0.8% calcium is sufficient for reproducing and growing budgerigars, the actual requirements may be lower. As clinical signs of marginal deficiency can take many years to develop it is prudent to ensure adequate availability.

Inadequate calcium in the diet of females during egg laying may result in thin or soft shelled eggs, although calcium deficiency is not the only cause of soft shells. Egg binding may also be partly due to a calcium deficiency since it responds well to an intramuscular injection of calcium gluconate (Harrison and Harrison, 1986). However as egg binding can occur in birds on a supplemented diet it is clearly a multifactorial condition.

Phosphorus is usually abundant in seeds but only 30–40% is present as non-phytate and can therefore be considered as available to the birds. If phosphorus is lacking in the diet, requirements for nervous tissue, cellular levels and enzyme formation have priority over bone formation. A partial deficiency will, therefore, always have an effect on the skeleton.

Trace Minerals

The requirements of many trace minerals can often be met from concentrations naturally present in feeds. Soils vary, however, in their content of trace minerals and therefore seeds from some geographic areas may be marginal or deficient in some elements. Supplementation may therefore be necessary to ensure an adequate intake. In addition, interactions occur between different trace minerals and can result in bioavailability problems (e.g. copper and molybdenum; calcium and zinc or manganese or iodine; mercury and selenium). Consequently, any dietary supplementation should be conducted accounting for these interactions, otherwise excessive concentrations of one mineral may result in a deficiency of another.

Minerals are also used in birds as colour pigments in plumage and skin. Copper, for example, occurs in turacin, a pigment of feathers conferring a blue colour. Hypopigmentation of feathers may also occur in birds deficient in iron.

Iodine deficiency has been frequently seen in budgerigars maintained on unsupplemented seed-based diets (Blackmore, 1963). It results in a decreased production of thyroid hormones which results in enlargement of the thyroid gland (goitre). This in turn can result in respiratory difficulties, regurgitation of food and a lack of activity. Symptoms of deficiency are seen more frequently in non-coastal areas where drinking water contains lower levels of natural iodine. Blackmore prevented thyroid dysplasia in budgerigars by supplementing the diet with 2 μg iodine twice weekly or adding 1% cod-liver oil (which contains high levels of iodine) to the standard diet.

Certain foods may contain substances which are goitrogenic. For example arachidoside in ground nuts will interfere with iodine utilisation (Guthrie, 1975). High levels of calcium can also exacerbate iodine deficiency by reducing absorption (Taylor, 1954; Ryan, 1991), therefore additional iodine may need to be provided for birds given *ad libitum* access to a calcium source.

Manganese deficiency in poultry results in perosis (slipped tendon). The hock joints become swollen and flattened and the Achilles tendon slips from its normal position. Perosis has been seen in ratites and cranes but not in passerines or psittacines (Dolensek and Bruning, 1978). This is probably due to their shorter limbs and lighter weight compared with ground-living species. However, although passerines and psittacines do not show clinical signs of deficiency it is not necessarily true that their requirements are different. Since low levels of manganese are found naturally in seeds, supplementation is probably prudent.

Zinc is generally abundant in most foods and particularly in the germ of seeds. However zinc from plant origin is less available than zinc from animal origin due to the formation of insoluble zinc–phytate complexes (Li and Valle, 1980). Calcium further decreases zinc availability by forming the more insoluble mixed zinc–calcium–phytate complex. The signs of marginal or chronic zinc deficiency include anorexia, poor wound healing, skin problems and scaly feet and poor reproductive capabilities. Excess zinc can also be harmful, inhibiting growth and causing anaemia. Very high levels of zinc are toxic and there is an increasing frequency of reported cases of birds dying from zinc toxicity due to the ingestion of zinc oxide from new galvanised mesh used in aviaries (Doneley, 1992).

Vitamins

Vitamin A

Vitamin A deficiency (hypovitaminosis A) is commonly seen in pet psittacines and most frequently in African greys and amazons. Clinical signs include increased susceptibility to respiratory infections, kidney disease, oral abscesses, decreased hatchability of eggs and high hatching mortality (Price, 1988). Birds are able to convert beta-carotene to vitamin A but most seeds are deficient in both of these. The majority of vitamin A deficiencies are caused by inadequate diets and particu-

FOODSTUFFS CONTAINING HIGH LEVELS OF VITAMIN A OR PRECURSORS	
Foodstuff	IU/100g as is
Beet greens	6,000
Carrots	11,000
Dandelion greens	14,000
Spinach	8,000
Sweet potatoes	9,000
Turnip greens	7,500
Dried red peppers	77,000
Beef liver	45,000
Egg yolks	3,000

TABLE 6.5: Foodstuffs containing high levels of vitamin A or its precursors. (Data taken from Pitts, 1983.)

larly, in parrots, due to overfeeding of sunflower seeds. Cod-liver oil is a good source of vitamin A, as well as iodine and is often used by breeders as a supplement. However oversupplementation can result in toxic doses of vitamin A and so care should be taken in evaluating the amount offered. Table 6.5 gives the vitamin A content of suitable foodstuffs for cage birds. Clearly some are more suitable as dietary sources than others if packeted supplemented diets are not used.

Vitamin D

Vitamin D_2 (ergosterol) is only about 1/30th as effective for birds as vitamin D_3 (cholecalciferol). Vitamin D is essential for adequate absorption of calcium from the digestive tract and for the interchange of calcium from bone. Inadequate supplies of dietary vitamin D, or access to ultra-violet light for synthesis of vitamin D from steroids in the skin, results in metabolic bone disease. Symptoms of calcium deficiency, such as the bones and the beak failing to calcify during growth and demineralisation of bones in adults are seen. Feathering may also be affected and with certain feather colours an abnormal blackening may develop. During reproduction a lack of vitamin D will result in eggs with thin shells. In chickens hatching success is reduced and embryos have malformed or only partially formed beaks. Most seeds, especially those ripened by the sun, contain high levels of vitamin D. Additional sources include fish-liver oils and eggs. It is probably only necessary to provide additional vitamin D to pet birds with little exposure to sunlight.

Vitamin E

Vitamin E deficiency in poultry results in encephalomania, exudative diathesis or muscular dystrophy. The signs of encephalomania are unco-ordinated movement, head retraction and outstretched legs and flexed toes. Exudative diathesis is caused by an increase in capillary permeability resulting in an accumulation of haemoglobin derived exudates under the skin over the breast. These exudates are a characteristic green colour. Nutritional muscular dystrophy, particularly of the breast, occurs when there is a combined deficiency of both vitamin E and selenium. Selenium can prevent both exudative diathesis and muscular dystrophy when vitamin E is deficient. Cockatiels with varying levels of paralysis have responded to a supplementation with vitamin E and selenium (Harrison and Harrison, 1986). Whole grains contain adequate levels of vitamin E and so deficiencies are rare on normal seed-based diets. Alfalfa and germ seed oils contain particularly high levels of vitamin E and can therefore be used as additional sources if necessary.

Vitamin K

Vitamin K is found in most green leaves and grasses. In addition, intestinal bacteria synthesise vitamin K which can then be absorbed by the bird. Only when antibiotics are being used will this non-dietary source not be available. Consequently deficiency is rarely seen (Altman 1978).

B-complex Vitamins

There have been no published studies on the requirements of companion bird species for B group vitamins. Thus our knowledge of requirements is based on work conducted on poultry and on the signs seen when they are given deficient diets.

Thiamin (B$_1$) is found in grain, although certain fungal infections of stored grains can reduce levels. Deficiency can also occur if birds become infected with specific intestinal bacteria possessing thiaminase which destroys dietary thiamin before absorption by the bird. Signs of deficiency include appetite loss, general weakness and polyneuritis. The latter develops rapidly in young chicks with lethargy, head tremors and, more progressively, retraction of the head over the back (Gries and Scott, 1972a).

Riboflavin (B$_2$)-deficient diets result in retardation of growth and curled-toe paralysis in chicks. Spontaneous recovery can occur in

cases of marginal deficiency and early stages are curable. A moderate deficiency can result in embryonic mortality in the middle of incubation and low hatchability (Peterson *et al.*, 1949).

Pyridoxine (B$_6$) and related compounds (pyridoxal and pyridoxamine) are widespread in seed diets and so deficiencies are rarely seen.

Biotin is also abundant in grains and so deficiencies of this vitamin are unlikely to occur in birds fed normal diets. In birds fed experimental diets deficient in biotin specific signs include dermatitis of the feet and lesions around the mandibles and eyes (Gries and Scott, 1972b). It is not clear whether birds are able to utilise biotin synthesised by intestinal microflora.

Vitamin B$_{12}$ is not found in plants but is present in meat, milk and yeast. Birds obtain some vitamin B$_{12}$ by absorption of the vitamin synthesised by bacterial microflora in the gut, or by the ingestion of faeces (coprophagia). A deficiency results in anaemia and gizzard erosion and has deleterious effects on skin and feathering. Chicks which hatch without adequate reserves from the egg may have a high rate of mortality.

Niacin (nicotinic acid) deficiency results in retarded growth, poor feathering and inflammation of the tongue and mouth. In chicks (as in the dog) this condition is known as 'black tongue' and increasing inflammation of the tongue and oesophagus result in reduced food consumption (Gries and Scott, 1972a). Most seeds contain sufficient levels of niacin except for maize. Since maize is also low in tryptophan, a potential precursor for niacin, supplemental sources must be provided in diets high in maize.

Pantothenic acid is important for growth and feather integrity in chicks. Signs of deficiency in chicks include retarded growth, crusty scabs at the corner of the mouth and the vent and dermatitis of the toes (in biotin deficiency it is primarily the foot pads which are affected). Seeds, such as wheat and oats, provide adequate amounts.

Choline is found in many foods and is present in sufficient quantities in seeds to prevent deficiency. Signs of deficiency in poultry include growth retardation and leg abnormalities. Adult birds are probably able to synthesise sufficient amounts to meet their requirements but dietary sources are required by growing and reproducing birds. As a complicating factor the requirement may be influenced by the level of vitamin B_{12} in the diet (Schaefer *et al.*, 1949).

Folic acid or folacin is so named because it is found in plant foliage. Birds deficient in folic acid exhibit retarded growth, poor plumage, loss of colour pigment from feathers and characteristic anaemia. The specific characteristic of folic acid deficiency anaemia is a reduction in the number of red blood cells which themselves become enlarged, malformed and contain more haemoglobin. Adult deficient birds may also show increased nervousness and a characteristic cervical paralysis in which the neck is stiff and extended.

Vitamin C

In general birds do not have a dietary requirement for vitamin C although there are known exceptions in the red-vented bulbul (*Pycnonotus cafer*) from Asia and one species of sunbird (*Aethopyga siparaja*) (Chaudhuri and Chatterjee, 1969). There is, however, increasing anecdotal evidence to suggest that supplemental vitamin C may be beneficial in times of physiological stress, such as reproduction, moulting or growth, although there are no scientific data to support this.

Conclusion

It is clear that there is still considerable scope for improving our knowledge of the nutritional needs of each species of companion bird, including aspects of feeding behaviour, energy, digestibility and nutrient content. In order to meet the nutritional needs of birds kept as companions one of the most effective policies is to mimic, as much as possible, the natural feeding ecology of the birds. This ensures that both the behavioural and nutritional dietary needs of the birds are fulfilled. In practice it is not necessarily possible to replicate the wide variety of plant species that are normally ingested throughout the year and seed mixtures, that still provide variety, are a practical alternative. However it is essential that such seed mixtures are adequately supplemented to account for the bioavailability of the nutrients and the life stage of the bird being fed (Robben and Lumeij, 1989). By feeding correctly supplemented seed mixtures the nutritional needs of the bird can be met.

PLATE 1: The modern accommodation units for (a) dogs and (b) cats at WCPN.

PLATE 2: Labrador puppies at play in one of the grass paddocks adjacent to the dog units. The puppies pictured here are aged from about 9 to 16 weeks.

PLATE 3: Two of the dog breeds at WCPN—Irish Wolfhound and Short-haired Dachshund—which illustrate the wide range of body size and type within this species.

PLATE 4: A dual feeder—one of the techniques which can be used to compare the palatability of different fish foods.

PLATE 5: A 7-day-old budgerigar chick. The eyes have opened and feathers are just starting to appear.

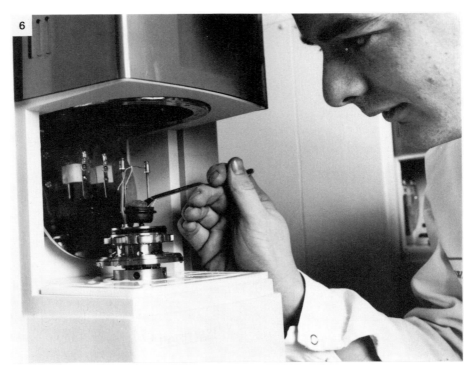

PLATE 6: A sample being loaded in to the bomb calorimeter for determination of gross energy content. The sample size is generally 1–2 g.

PLATE 7: The new cat metabolism accommodation at WCPN, showing the spacious lodges for housing individual adult cats.

CHAPTER 7

Nutrition of Ornamental Fish

MARINUS C. PANNEVIS

Introduction

The increased popularity of fish keeping worldwide over the last 20 years has resulted in an increased demand for scientific knowledge on what and how to feed and how to care for ornamental fish in general. Today a larger number of pet fish is kept worldwide than all the other pets added together. In the USA alone, over 100 million fish are kept as pets.

Ornamental fish nutrition is perhaps one of the least explored areas of pet nutrition, and it is regrettable that this subject has so often been overlooked in respect to the work done for mammalian pets. The health and well-being of ornamental fish is largely dependent on the pet owners to supply the appropriate quantitative and qualitative dietary requirements. Although ornamental fish have been kept for over 1000 years (in Europe since 1600), the hobby has only become widespread in the last 40–50 years with increased fish breeding and imports. The availability of fish foods and aquatic hardware (tanks, filters, heaters and water treatments) have also made this form of pet ownership more accessible.

Ornamental fish have traditionally been fed on live foods (tubifex worms, daphnia, earthworms, even other ornamental fish such as goldfish and guppies) due to their high palatability. In the past most pet fish were caught in the wild (and then kept as pets), so that live food elicited a stronger feeding response. Live food is not always available throughout the year and not suitable for all species of ornamental fish (i.e. herbivores). Moreover, live foods can be nutritionally deficient if the food organisms have been starved themselves (tubifex kept for days in a pet shop without food), or have received deficient diets before being used as a food source. Live foods are also profound transmitters of diseases (parasitic, bacterial and viral). Live foods have more recently been marketed as dried or freeze-dried products (insect larva, shrimp species) in order to overcome year-round supply problems and to reduce some of the disease risks.

The introduction of freeze-dried products has not been without problems. The reduced vitamin levels, depending on the drying method, cause nutritional stress as does the oxidation of polyunsaturated fatty acids in sun-dried aquatic diets (krill, gammarus, etc). These products therefore have limited use as complete diets.

In the 1950s manufactured fish food made its first significant entry into the market. The formulations were initially based on using large varieties of exotic raw materials to reflect 'natural diets'. Development of the flake as the preferred way to feed ornamental fish is still used today. The classical advantage of flakes is that they can be stored easily and sprinkled on the water to float, or be submerged for a few seconds to sink slowly, depending on the density of the formulation. The flake is thus an ideal form to attract both surface, mid-water and bottom-feeding fish species. Good flakes have both sinking and floating properties to give all occupants of the tank equal opportunity to feed.

The first real impact in fish nutrition research came from the growth of the aquaculture industry in the 1970s. Aquaculture research has, however, been based on a small number of food fish species, often kept under totally different husbandry conditions as compared with ornamental species.

Over the last 20 years more than 68 nutrient requirements for a number of 'aquaculture' fish species have been identified. However the documented nutrient requirements of these species are not necessarily directly relevant to the nutrient requirements of ornamental fish. This has been one of the key reasons for establishing the Waltham Aquacentre, a centre that studies purely the nutrient requirements of ornamental fish, their feeding behaviour and care.

Ornamental fish include a large number of species and breeds from a wide variety of geographical habitats (e.g. Amazon, African Lakes, South East Asia, Japan, China, Pacific Ocean Reefs). Fish are the largest group of vertebrate species on earth: there are approximately 30,000 different species, although this number is currently dwindling under severe environmental pressure (e.g. the Amazon and Coral Reef destruction). Of all the fish species, 4000–5000 are being kept as pet fish worldwide. Only a few hundred of these are very popular and commonly kept by a large number of fish hobbyists while specialist-hobbyists and zoos are often involved with the rare species.

The ornamental fish nutritionist, whether in food manufacturing or responsible for public aquaria, is confronted with the problem of providing diets for many fish species for which there are no specific individual nutrient requirements. In an aquarium environment it is practically impossible to feed specific diets as different species are kept together. For example, foods that may have been formulated for carnivores will also be eaten by herbivores. In general, it is therefore more logical to approach the nutritional classification for one *group* of pet fish rather than for a single species.

The general nutritional classification of ornamental fish is based on biotic (i.e. physiology, life stage or feeding behaviour) and abiotic factors (i.e. environmental temperature, salinity). As fish are poikilotherms (they maintain a body temperature equal to the water temperature), the most important abiotic nutritional classification is that of temperature origin. One can distinguish between pet fish originating from temperate or tropical regions or even arctic regions although the latter are very rarely kept as pets. The tropical species are much more abundant than the temperate species, possibly as a result of a larger diversification in the tropical environments and subsequently wider arrays of fish species with numerous feed specialisations. The second abiotic nutritional classification is one of environmental salinity and water hardness. The obvious distinction is between sea water and fresh water, but many intermediates are common such as soft/hard water or brackish. The most important biotic nutritional classifications are those of herbivorous, omnivorous or carnivorous ornamental fish. The combination of these three main classification groups (e.g. a tropical freshwater omnivorous fish) results in more than 18 distinguishable nutritional groups of ornamental fish. There is often overlap between the groups. For example, in natural habitats in a temperate climate (e.g. European garden ponds) you will find herbivorous fish becoming detrivores (fish that retain food from the bottom surface of a habitat) in the winter, eating small invertebrates, as plankton and aquatic vegetation becomes scarce.

Of all the distinguishable nutritional groups of ornamental fish the most important are the temperate freshwater omnivores (koi carp, goldfish) and the tropical freshwater omnivores (most tropical community fish). Other less common groups of ornamental fish fall within the classification of tropical marine herbivores/omnivores/carnivores (coral reef species) and the tropical freshwater carnivores/herbivores (Cichlidae).

As most tropical pet fish (marine and freshwater) are kept in community tanks, a compromise is needed to accommodate individual nutrient requirements.

Most carnivorous pet fish are in fact quite peaceful community fish as they are specialised in eating insect larvae or invertebrates rather than other fish. Piscivorous fish (fish-eating fish, such as piranha) should not be kept with other non-specifics for obvious ethical reasons! It is very regrettable that feeding live fish to piscivorous fish is still practised despite the fact that these species can be trained to feed on artificial diets.

Digestion

Most fish engulf and swallow their food whole, although there are some exceptions. To be successful in handling artificial foods the fish has to develop an integrated approach (dentition, structural support and behaviour) to the diet. The anatomy of the digestive tract of ornamental fish is very diverse as one would expect with over 4000 ornamental fish consisting of herbivores, carnivores and omnivores. Most fish do not have teeth, however some of the marine coral reef species have a well developed dentition adapted to crushing live coral to release the internal nutritious parts for digestion whilst excreting the sand undigested. The marine coral butterfly fish (Chaetodontidae spp.) has a dentition that is specially adapted to eat living coral. The small, terminal mouth contains fine incisiform protruding teeth for biting off individual polyps. The Scaridae or the marine coral parrot fish has also a strong protruding beak but fused teeth which can be used to scrape polyps rather than break them off. The marine coral trigger fish (Balistidae) has a very strong protruding beak that can be used to break off and ingest ends of coral heads. However most fish do not possess these specialised food processing mechanisms and food is thus transported straight into the oesophagus. A number of herbivores, such as the koi carp, have no defined stomach and consequently lack the pepsin and acid hydrolysis stage of digestion. It has been hypothesised that the lack of a stomach in herbivorous fish could be an evolutionary response to low protein intake. Alternatively, as many herbivores are freshwater fish species, there may be insufficient chloride to produce sufficient hydrochloric acid for stomach acid hydrolysis. Other fish species, often carnivores, do have a stomach and are able to carry out acid hydrolysis of dietary protein. Marine ornamental fish also have mechanisms to counteract the uptake of alkaline water (pH 8.5) particulary whilst acid hydrolysis is required. A number of theories have been postulated to explain the mechanisms involved in this process like alternate drinking and feeding, contact digestion in which food is 'fenced off' from sea water, and the anatomic explanation for a Y-shaped stomach in which sea water could pass through and food would be retained for digestion. The stomach could in this case also cater for the dilution process of sea water to reach isotonic concentrations equal to blood osmolarity. In fish with no stomach the bile and pancreatic excretions enter the intestine at the point where the oesophagus joins the ileum. Herbivorous fish possess a larger ratio of intestinal tract to body length, this ratio ranges from under 0.5 to over 15. In carnivorous fish the ratio varies between 0.2 and 2.5, with omnivores having an intemediate ratio between carnivores and herbivores.

It is thus not surprising that the digestion of the major nutrients is not uniformly efficient for all species of ornamental fish. There are a large number of factors that determine the digestibility of protein, fat and carbohydrate. Especially with the primitive digestive tract of the herbivores, the feeding rate is a driving force for gut transit time. That means that as the fish eat more, the speed at which the food passes through the gut increases and the digestion of the food is much reduced. If a herbivore is fed on unlimited palatable plant material the faeces will show that the fraction exposed to the gut wall has been partially digested but the inner section of the faeces is still undigested and green in colour.

The evaluation of digestibility for ornamental fish at the Waltham Aquacentre takes place using specially designed tanks that allow the continuous collection of faeces from even the smallest ornamental fish species. Diets are routinely fortified with 0.5% undigestible chromic oxide as a digestibility marker to evaluate protein, fat, carbohydrate and total energy digestibility. Another method often used is total weight gain per gram food supplied. Our work has shown that a four-fold difference exists between diets in terms of this conversion factor, indicating that some ornamental fish foods are very poorly utilised.

Energy Requirement

As ornamental fish are poikilotherms, this is reflected in their maintenance energy requirement which increases with increasing water temperature. For koi carp the daily maintenance energy requirement more than doubles from 10 to 20°C. This is an important consideration when planning a feeding strategy for pond fish during the different seasons.

From late autumn to early spring, pond fish virtually stop feeding when water temperatures reach 5°C and lower. Under these conditions their metabolism is reduced to a hibernation level and the fish survive by utilising their fat and protein deposits until water temperatures rise in mid-spring. In contrast to mammals, very little glycogen is mobilised as an energy source in fish. It is vital to prepare pond fish for winter hibernation by feeding them food with higher levels of digestible energy in the late summer and autumn. Nevertheless even in the summer, and for tropical fish which are kept at constant temperatures year round (26°C), their energy requirement is far lower than that of homeothermic mammalian pets. The energy requirement of ornamental fish is typically between 5–10% of

that of the mammalian dog in terms of $J/g^{0.75}/$ day. Koi carp at 20°C require on average of only 1.4% of the energy needs of a dog after correcting for metabolic body size (Table 7.1). In spring and autumn and during poor summer weather the requirement of koi are even lower as compared with the mammalian dog.

There are other reasons, some related to the physiology and behaviour of poikilotherms that explain their relatively low energy requirement. Firstly, the metabolic pathway of protein breakdown results in the production of ammonia which can be excreted over the gills by diffusion and does not need to be concentrated and transformed into urea as in most mammals. Although some urea is formed it is insignificant compared with their excretion of ammonia. Fish also require less energy because of the efficient way in which they move through water as compared with our mammalian pets on land. This latter factor is of course heavily dependent on the water current generated by air-pumps and filters, which may turn over 200–800 l/hr through a 150 litre aquarium.

The main sources of metabolic energy for fish are from lipids and protein although some species (mainly herbivores and some omnivores) can utilise significant amounts of carbohydrate. The metabolisable energy utilisation of carp, a close relative of ornamental koi, was more than 2.5 times more efficient from fat than from wheat flour (85% carbohydrate) or fish meal (70% protein). Considering that carp are relatively efficient in the utilisation of carbohydrate as compared with other species, it is not surprising that fat is a more favoured energy substrate for tropical community fish. The metabolic breakdown products of fat metabolism are carbon dioxide and water whereas protein also liberates ammonia. The ammonia from protein catabolism will become a nutrient for bacterial and algal proliferation. Even after full bacterial nitrification, the accumulation of nitrate becomes a stress factor for fish. Special attention has thus to be made to formulate diets for ornamental fish with enough non-protein energy so that a minimal amount of protein is

COMPARISON OF ENERGY REQUIREMENT OF THE DOG AND ORNAMENTAL FISH			
Species	Bodyweight (g)	Allometric energy constant*	
		20°C	26°C
Dog	20,000	2900–3000[†]	
Koi carp	20–40	31–53	—
Goldfish	5	80	240
Moonlight gourami	8	—	150
Neon tetra	0.2	—	210

TABLE 7.1: Comparison of energy requirements of the dog and the ornamental fish. * Related to body weight$^{0.75}$ (g) and energy requirement (J/day). † Depends on activity

broken down. There is, however, limited knowledge to date as to the extent to which dietary fat can be fed as a protein-sparing energy substrate while at the same time avoiding excessive fat deposition and changing body composition.

It is also important to appreciate that energy utilisation is dependent on the feeding strategy we employ when feeding small ornamental fish (e.g. neon tetras; *Paracheiron innesi*) or fish that do not have a stomach, in which case no capacity exists to take a large meal and digest this over time. Small fish and fish without a stomach need frequent small meals that can be utilised efficiently. This has been demonstrated at the Waltham Aquacentre with neon tetras (*Paracheiron innesi*) with a average body weight of 0.2 g which were fed the same quantity of total food per day but at different frequencies. It was found that 62% higher growth rates were achieved when food was fed in three small portions compared with one large meal. However when food is offered in very small portions, growth and utilisation become very inefficient due to non-recognition of food by the fish. Excess food will be trapped in the gravel of the tank or in the filter and is then unaccessible for most fish. This emphasises a big difference between feeding other pets where food is presented in one defined area as compared with fish where food is available, not concentrated in one place, but scattered in a space up to many thousand times the fish's own body size. Small species of fish with a high metabolic rate such as the neon tetra can lose up to 3.5% of their body weight each day during starvation. A successful feeding regime is characterised, therefore, by optimal utilisation of food by all species while avoiding waste through non-recognition of food, or excessive feeding.

Protein and Amino Acid Requirements

Protein forms the largest quantity of dry matter in fish and a continuous dietary supply is needed to maintain body protein turnover and growth. As with most mammalian pets, fish require 10 essential amino acids in their diet: arginine, histidine, isoleucine, leucine, lysine, methionine, phenylalanine, threonine, tryptophan and valine. The biological value of dietary protein for ornamental fish is determined by the quantity of these essential amino acids in the diet in comparison to the individual dietary requirements of the species involved. Where there are a number of species kept in a community tank the level of each essential amino acid should be included at the level necessary for the species with the highest requirement.

Unlike fish in the wild, which live in large water systems, captive ornamental fish need to utilise their dietary protein with the utmost efficiency as metabolic breakdown products of protein (mainly ammonia and only a little urea) will directly pollute their environment. Metabolic waste accumulation, especially of nitrogen compounds, is a serious threat to the health of ornamental fish in an aquarium and is often overlooked by feed manufacturers and aquarists. Generally, most fish food

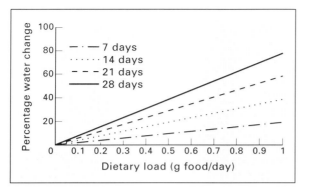

FIG 7.1: Interaction between feed gift and maintenance of aquaria.

larval and juvenile fish, one requires high protein levels (45–55%) to sustain rapid growth, but generally this should always be accompanied by good filtration systems and frequent water changes with nitrate-free water to reduce the nitrate concentration which accumulates after protein breakdown and bacterial nitrification. In Fig. 7.1 the relationship between the level of feeding and water changes is demonstrated for a 100 l tank. For example feeding 0.2 g of food per day requires a 3.8% water change every week or 7.5% every fortnight or up to 15% per month.

contains 30–55% protein, however on some occasions a closer examination reveals the use of plant protein with low levels of essential amino acids in the formulation. When plant protein is used in formulated fish foods the lack of some essential amino acids will result in the protein being broken down as an energy substrate which loads the environment with nitrogen breakdown products resulting in rapid algal growth and stressful water conditions for fish. So although protein content is important, the biological value of the protein is the most critical factor determining its suitability as an ornamental fish food. The biological value in terms of available lysine and other amino acids will be influenced by the carbohydrate content of the diet and the temperature to which both nutrients are exposed during manufacture.

Proteins from animal sources (fish meal) are a prefered ingredient in all fish diet as they can be utilised efficiently by even the strictest herbivores. Although these fish are adapted to utilise some complex carbohydrates and so release the enclosed proteins of the plant, the process itself is not efficient and therefore not suitable for an aquarium environment, where non-utilised ingredients and metabolic breakdown products will accumulate and cause environmental stress to fish. Although protein and free amino acids are highly palatable to fish, when maintenance feeding is required protein levels around 50% are unnecessarily high for omnivores and herbivores. In some cases, such as feeding

Fat and Essential Fatty Acids

Dietary fat is the most important source of energy for ornamental fish. It is vitally important to include enough dietary fat in order to reduce dietary protein used as an energy substrate (protein sparing effect) which results in high nitrate concentrations in the aquarium. The metabolic waste products of fat breakdown are harmless carbon dioxide and water. As fish have a generally low energy demand, care should always be taken to reduce fat levels to less than 15% to avoid deposition of fat which can cause fatty liver degeneration in extreme cases. Dietary fat is also an important vehicle for the uptake of the fat soluble vitamins (A, D, E and K).

Fish have an essential requirement for certain n-3 and n-6 polyunsaturated fatty acids which they cannot synthesise themselves. Essential fatty acid deficiency in fish is best prevented by supplementing linolenic (n-3) and linoleic (n-6) acids. Fish in general require fatty acids of longer chain length and higher degree of desaturation than the dog and cat. The reason for this lies in the fact that fish have a lower body temperature than mammals and these fatty acids with low melting points are needed to support cell membrane flexibility. The melting point of the fat is correlated with the chain length and the degree of unsaturation of the component fatty acids. Generally the lower the melting point of the fat the higher the unsaturation and the higher the digestibility for the fish.

Depending on environmental temperature, salinity and life stage, the essential fatty acids need to be supplemented in a different ratio of n-3 and n-6 fatty acids in the diet. Between 5 and 10°C the n-3/n-6 ratio is close to 2, while between 15 and 20°C it reduces to 0.5.

The high requirement for n-3 fatty acids at low temperatures should receive proper attention in formulating diets for cold water fish in outside ponds and cold water aquaria. It has also been suggested that salinity increases the requirement for n-3, but this could also be related to a lower temperature effect of sea water compared with more shallow, tropical, fresh water lakes. Dietary deficiency of these essential fatty acids causes increased mortality, erosion of the fins and, most profound, 'shock syndrome' in which the fish cannot control the nerve responses to its muscles and will swim erratically into the side of the tank or pond.

Recently more emphasis has been laid on the possible requirement for HUFAs (highly unsaturated fatty acids), especially the role of docosahexaenoic acid (DHA) for larval stages of fish. As larval fish require more fat in their diet, special care is required to avoid a lipid film occurring on the water/air interface of the tank which can prevent oxygen from dissolving in water and make it impossible for the larvae to fill their swim bladder with air just after hatching. Larvae cannot control their buoyancy so if they have no access to atmospheric air they will subsequently die of swimming exhaustion. There is also increasing evidence that larval fish utilise certain enzymes, present in their prey, to facilitate digestion of these same prey in their simple digestive tract.

Unsaturated dietary lipids are very liable to oxidation while stored with atmospheric oxygen or at high temperatures and in direct (sun)light. The oxidation metabolites can be toxic to fish and proper care is required to avoid this breakdown process by the use of antioxidants in the feed. As with other species, for fish there seems to be a link between the dietary PUFA content and the vitamin E (a natural antioxidant) requirement (see Chapter 2). When saturated fat is fed to fish at low temperatures it may solidify in the gut and block the intestinal canal, so this practice is not recommended.

Carbohydrate

Carbohydrate is not an essential nutrient for fish. Most herbivorous fish have developed with physiological adaptations, such as a long intestine/body length ratio, that promote microbial digestion of carbohydrate. Carbohydrate is not as abundant below the water surface as it is in a terrestrial habitat. Carbohydrate is most abundant in macrophytes (i.e. water plants, weeds, algae) rather than in aquatic animals and is thus only available where plants can live, that is shallow waters that allow direct light to penetrate the water for photosynthesis. Some herbivorous fish, and also the omnivorous-herbivore goldfish and koi carp, use the microflora in their hind gut to digest the complex carbohydrates (i.e. dietary fibre) for which they themselves do not possess the enzymes. It is thus not surprising that carbohydrate digestion can vary greatly between species of fish, and even between fish of one species depending on environmental conditions. Work carried out at the Waltham Aquacentre showed that carbohydrate digestibility can vary from 70% in goldfish (*Carassius auratus*) to as low as 50% for moonlight gourami (*Trichogaster microlepis*). This means that for the moonlight gourami 50% of the carbohydrate is excreted as non-digestible matter. This non-digested carbohydrate is a potential substrate for bacterial proliferation in the aquaria which is undesirable.

Minerals

Ornamental fish require minerals for normal body metabolism such as bone and haemoglobin formation, pH regulation, and for hormone and enzyme function. Unlike terrestrial pets, ornamental fish can absorb some water-soluble minerals so in theory they can be independent of these dietary minerals. However as the mineral composition of water is very variable it is recommended that dietary supplements are used to satisfy the full nutritional requirement of fish kept in aquaria containing mineral deficient water.

Calcium and Phosphorus

Gills are the most important regulators of calcium balance in fish. Dietary calcium requirements can be very low in waters rich in calcium as fish can transport calcium from the water over the gills into their blood. It remains questionable as to how much water-soluble calcium in an aquarium can be utilised, particularly when little fresh water is introduced into the tank and water sources are low in dissolved salts. The calcium is stored in bones and scales, however bone does not play a major role in calcium regulation. The digestion of calcium in the gut seems to be better in fresh water than in marine fish species owing to interactions with sea water minerals (magnesium, strontium, zinc and copper). Supplying a source of dietary phosphorus is very important as water contains only low levels of this mineral. It has been suggested that phosphorus is taken up over the gills. Most fish are unable to utilise phosphorus in the form of phytate because they do not synthesise phytase.

Magnesium

Magnesium is either actively taken up over the gills or digested through dietary uptake in the gut. Magnesium is stored in the bones of fish. In a marine environment the dietary supplementation of magnesium is most probably unnecessary.

Sodium, Potassium and Chloride

Sodium and potassium are the most important cations while chloride is the most important anion for fish. They have a key role in maintaining pH control and normal nerve functioning. Ornamental fish can obtain sodium, potassium and chloride both from water and the diet. Because they can utilise the water-soluble minerals it has been difficult to ascertain dietary requirements. Most ingredients in fish foods do contain enough of these minerals and it is subsequently very rare to find any deficiency symptoms especially for marine fish as sea water normally contains adequate amounts of all three minerals.

Iron

Iron is considered to be taken up mainly via the intestine although it has been shown to pass across the gills. High iron levels in the water (> 200 ppm) have a negative effect on the gills and can cause poor oxygen transfer from water to blood.

Copper

Copper is closely involved in iron metabolism and a number of enzyme functions. Although copper is an essential nutrient, the tolerance levels are very close. The normal requirement is around 2–3 ppm, whereas concentrations over 15 ppm can cause growth depression.

Iodine

Iodine is a key component in the formation of the growth hormones thyroxine and triiodothyronine. Iodine can be directly absorbed from the water by fish, but iodine levels in seawater systems are rapidly depleted. The dietary requirement is estimated between 1 and 5 ppm.

Zinc

Zinc is active in many enzyme systems. It seems that zinc is best supplied through the diet rather than through the water. Dietary zinc deficiency causes cataracts, poor appetite and growth.

Manganese

Manganese plays a role in the enzymatic synthesis of urea from ammonia, and is thus involved in the protein catabolism of fish. Deficiency of manganese can cause abnormal tail growth, poor growth, anorexia, and loss of equilibrium.

Selenium

Selenium is a very important mineral that plays a role in the enzyme that reduces the risk of peroxides formed by the oxidation of polyunsaturated fatty acids in the fish. As with copper the tolerance range is small with an upper toxic limit of 15 ppm.

Other minerals

Fluoride plays an important role in the marine food chain and has a function in bone formation. Fish seem to have a high tolerance level for fluoride in the water. There is absolutely no evidence that fluoride added to drinking water will harm ornamental fish. The requirement of cobalt is very low but is essential in vitamin B_{12} synthesis. Cobalt deficiency can thus cause avitaminosis B_{12} symptoms.

Vitamins

Ornamental fish have roughly the same qualitative requirements for vitamins as mammalian pets. However, while quantitative vitamin requirements have been established for a number of aquaculture fish species there is no information for any of the ornamentals. It has to be assumed that the requirement for ornamental species is at least equal to the aquaculture species of which we have species-specific requirements. As ornamental fish have very limited access to natural food the vitamins must be supplemented through the diet. Only 4 out of 7 commercial ornamental fish foods guarantee the vitamin composition until the sell by date. This is not surprising as some fish food may be in the retail chain for up to three years after the date of manufacturing and subsequently deficiencies in vitamins may be common. As high temperatures are required to manufacture food flakes some losses already occur during manufacturing. This means that in order to overcome process and storage losses higher doses of vitamins need to be used in the premix formulation. Little work has been carried out to evaluate the effects of hyper- or hypovitaminosis on ornamental fish.

As fish foods are being administered in water, the rate of loss of water-soluble vitamins is very high. Depending on the solubility up to 90% of some of the B vitamins, and 65% of vitamin C, can be leached out of the food within 30 sec of being in contact with water. Of all flake foods tested the average time at which the first flake arrived at the bottom of a tank (30 cm deep) was 90 sec. For bottom-feeding fish species this is the first chance they have to gain access to the food, assuming other fish have not eaten the diet. Hypovitaminosis can thus be expected in this group. This problem is widespread and has been shown to be the case for more than 70% of the available commercial ornamental fish diets. Some marine fish species will drink sea water but are unlikely to retain even a fraction of the water-soluble vitamins once they have leached from the food. The nature of vitamin leaching and the fact that evaluation of vitamin intake is difficult make dietary requirement studies in most cases very unreliable.

The vitamin deficiency symptoms described below, while typical, have not been confirmed for all species of fish. Some symptoms may be the result of secondary infections related to the vitamin deficiency.

Vitamin A occurs as vitamin A_1 ($C_{20}H_{29}OH$) or vitamin A_2 ($C_{20}H_{27}OH$). While A_1 is only found in mammals and marine fish, the A_2 form is common in freshwater fish species. As a factor in the regeneration of rhodopsin it forms an important factor in vertebrate vision. It also has a function in calcium transport, reproduction and membrane integrity. Deficiency symptoms are anorexia, ascites, lens deformation, edema, exophthalmos (eye protrusion), fin erosion, poor growth, eye haemorrhage, kidney haemorrhage, skin haemorrhage and eye lesions.

Vitamin D, as D_3 (cholecalciferol), has a function in calcium absorption from the intestine. Although fish can absorb calcium from the water over the gills a requirement for this vitamin certainly exists in areas where fish are kept in soft water. Deficiency signs are poor growth, scoliosis (sideways curvature of the backbone) and white muscle tetany.

Vitamin E, or tocopherol, forms an essential function in the protection of unsaturated fatty acids against oxidation. The dietary level of vitamin E is therefore related to the level of dietary polyunsaturated fatty acids. Deficiency symptoms are anaemia, ascites, muscular dystrophy, oedema, exophthalmos, fatty liver, erythrocyte fragility and fragmentation, poor growth and reduced haematocrit.

Vitamin K has a function in the blood clotting cascade and is normally included in fish food as vitamin K_3, menadione. Deficiency symptoms are anaemia and slow blood clotting.

Thiamin plays a role during carbohydrate metabolism in fish. For carp, a close relative to the koi carp, a correlation has been found between thiamin requirement and carbohydrate requirement. Thiamin deficiency results in anorexia or poor growth, muscle atrophy, cataracts, convulsions, oedema, neurological disorder, equilibrium loss and skin colour darkening.

Riboflavin is involved in the electron transport system and plays a role in coenzymes of fish. Riboflavin deficiency results in anaemia, anorexia, ataxia, cataracts, dark skin coloration, fin erosion, poor feed utilisation and growth, eye haemorrhage, skin haemorrhage, eye lesion, photophopia, iris pigmentation and cornea vascularization.

Pyridoxine is an coenzyme involved in amino acid metabolism. The requirement seems to be dependent on the dietary protein level, although this has so far not been proven for omnivorous fish species. Deficiency symptoms are anaemia, anorexia, ataxia, dark skin coloration, convulsion, oedema, equilibrium loss, exophthalmos, rapid gasping, poor growth, irritability, nerve disorder, rapid rigor mortis, blue slime and erratic swimming.

Pantothenic acid plays a role in the formation of coenzyme A, which is involved in fatty acid oxidation and synthesis and a number of other enzymatic conversions. Deficiency signs are anorexia, ataxia, gill atrophy, gill exudated and clubbed, dermatitis, poor growth, skin haemorrhage and lesions, lethargy, liver necrosis, prostration and erratic swimming behaviour.

Niacin is involved in the coenzymes MAD and NADP. Fish seem to be unable to synthesise sufficient quantities of niacin from tryptophan, so that niacin is required in the diet. Deficiency of niacin results in anaemia, anorexia, edema, poor feed utilisation and growth, skin haemorrhage, colon and skin lesions, lethargy, photophobia and white muscle tetany.

Biotin plays a role in the biosynthesis of long chain fatty acids and purine and acts as an intermediate carrier of the carboxyl group during carboxylation and decarboxylation reactions. Deficiency symptoms are anorexia, muscle atrophy, dark skin coloration, convulsion, gill degeneration, fatty liver, poor feed utilisation, erythrocyte fragility, colon and skin lesions and blue slime.

Folate plays a role in amino acid metabolism. Deficiency signs include anaemia, anorexia, dark skin coloration, poor feed utilisation, fin fragility and lethargy.

Vitamin B_{12} plays a role in erythrocyte maturation, fatty acid metabolism and in the recycling of folic acid so that the deficiency signs of the two are similar. Other vitamin B_{12} deficiencies are erythrocytes fragmentation and a low haemoglobin value.

Choline has no coenzyme function but acts as a precursor of the neurotransmitter acetylcholine and functions as a methyl donor. Choline deficiency symptoms are fatty liver, poor feed utilisation and growth and kidney haemorrhage.

Inositol has also no known coenzyme function but acts as a growth factor in fish. The deficiency symptoms in fish are anaemia, anorexia, distended stomach, fatty liver, poor feed utilisation and growth, and skin lesions.

Vitamin C cannot be synthesised by fish unlike most mammals and birds. Ascorbic acid is a strong reducing agent and a cofactor in the hydroxylation of proline to hydroxyproline, a precursor of collagen. Owing to the low stability of ascorbic acid the vitamin is often included as a phosphate- or sulphate-stabilised vitamin C or as micro-encapsulated vitamin C. Not all the stabilised vitamin C varieties can be equally efficiently utilised by the different species of fish. It has been suggested that sulphate-stabilised vitamin C can be digested but has a subsequent lower bio-availability. Deficiency symptoms of vitamin C are anaemia, anorexia, ascites, abnormalities in the cartilage, low resistance to diseases, exophthalmos, poor growth, reduced haematocrit, low haemoglobin, gill haemorrhage, kidney haemorrhage, liver haemorrhage, skin haemorrhage, eye lesion, lethargy, lordosis (upward curvature of the backbone), prostration and scoliosis.

Pigmentation

Ornamental fish occur in a wide variety of colours (red, purple, blue, yellow, silver, green and in all shades and combinations), all of which are natural and form an important part of the social behaviour of fish. The basic building blocks for these colours in the skin (and flesh) are all derived from the diet. Most of the pigments are carotenoids, and more than 600 carotenoids have been isolated so far. Animals, in contrast to microorganisms and plants, are unable to synthesise carotenoids *de novo*. Modifications of carotenoids by biochemical pathways are species dependent. Although the carotenoids are typically yellow, orange to red, the complex carotenoids such as carotenoid lipoprotein, carotenoid proteins and other derivatives extend the colour range into blue, green and violet. Astaxanthin and canthaxanthin are the carotenoids most often used for pigmentation of fish. The way in which fish can metabolise the basic carotenoids such as beta-carotene is species dependent. Goldfish for instance cannot convert beta-carotene into astaxanthin to produce their orange skin colour.

Palatability and Feeding

Fish use a range of sensory mechanisms, such as optic, acoustic, olfactory and electro-organ discharge receptors to detect food. It depends on the species and the characteristics of the habitat in which the species evolved as to the form of sensory mechanism that is most important. Most catfish are equipped with barbels (sensory organs around the mouth area) which they use to search for chemical clues, as optic clues are often not appropriate for the dark, murky water conditions in which they live. Obviously the function of barbels in the clear water of an aquarium with gravel rather than a mud substrate is perhaps less important. Killifish, on the contrary, are greatly dependent on visual clues, although final food intake is dependent on chemical receptors. Piranha also feed on visual clues, avoiding food that resembles their own body shape. It is unclear how important optic factors are in determining the arousal of fish to initiate feeding. Contrary to popular belief, colour clues have been shown to be as important to fish as shape and contrast. There is also some evidence for species-specific colour preference. Besides colour, the contrast between food colour and its environment is an important trigger for food detection. Night-feeding fish are vulnerable to predation during the day and have fine non-visual ways to search for food whilst using the darkness to approach their food close to striking distance. If one has night feeders in the collection it is important to note that if food is available only during the day they may starve. In that case it is better to feed small amounts of food after turning the light off in the tank. Some species have developed electrosensory mechanisms which can provide information about the location, size, shape and even quality of the objects in the predators' immediate vicinity, either hidden below sand or mud. Sound is also an important clue to find food in the wild (e.g. insects) but it has become less important in an aquarium. Like mammals, fish distinguish between taste and olfactory signals. But unlike mammalian pets, ornamental fish have taste receptors scattered over their whole body and not just in the mouth. The gustatory response is different for each species of fish and can even be different for one species from different geographical locations. Work carried out at the Waltham Aquacentre has shown that ornamental fish can exhibit a very consistent preference for one diet over another when the two diets are fed simultaneously.

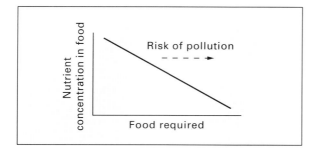

FIG 7.2: Relationship between food required and its nutrient content on water quality in the aquarium.

This technique is used to determine species-specific preference for diets. Protein and especially the hydrolysed free amino acids, betaine and other amino acid type molecules of less than 1000 molecular weight, are important chemical attractants for fish. Fish react to L-amino acids but not to D-amino acids. To operate as a feeding attractant a critical concentration of free amino acids needs to be reached. In general terms it has been found that herbivorous fish are attracted to some single free amino acids common to plant protein while carnivorous fish react to a combination of free amino acids most common to aquatic vertebrate and invertebrate tissues. Palatability in terms of rapid food recognition and uptake is important not only in terms of owner enjoyment of feeding their fish but also help to avoid water soluble nutrients, such as vitamins, leaching out and becoming unavailable to the fish. A large number of ornamental fish is still caught in natural habitats. It is not surprising that these fish do not instantly recognise flakes or pellets as food. It is therefore important that collectors and retailers initially feed live food, despite the additional disease risk. After a period of adaptation a wild fish will recognise flakes as food. It can also be helpful to house 'domesticated' fish in the same tank as new wild fish so that the latter can learn how to select food. It is known that fish living in schools find food faster than lone individuals.

Conclusion

The factors affecting the nutrition of ornamental fish are far more complex than that of feeding single species of other groups of pets. The three dimensional feeding environment, water-soluble vitamin leaching, inter-species competition and individual species palatability preferences are among the examples discussed in this chapter. Although fish can obtain enough of their nutrient requirement by increasing their intake of low nutrient concentrated diets, the repercussions on pollution of their own living environment make this an impractical feeding regime (Fig. 7.2). It is clear that nutrient-dense diets with high biological values and correct proportions of digestible non-protein energy are required for maintaining ornamental fish in a healthy condition.

CHAPTER 8

Nutrition of Horses

WILFRIED E. TIEGS and IVAN H. BURGER

Introduction and History

For a long time the horse has had a special relationship with man, but this has seen many changes. In the early part of this century the number of domestic horses showed a steady decline as they were gradually replaced in their main role as a provider of motive power. More recently (from around the 1960s) they have experienced a revival which has been mainly due to their increased use for leisure and sports. To this extent, therefore, the horse can be considered to be a true companion animal and merits a place in this book alongside the more 'traditional' species. The world population of horses is currently around 65 million.

Like the dog, the horse family includes a large number of different breeds (including ponies), and where reference is made to the horse this should be taken to refer to all of the species (*Equus caballus*). This chapter is aimed at being a brief overview of the key points of horse nutrition, highlighting special aspects of the subject, such as the influence of activity on nutrient requirements.

To explain the nutritional needs of horses, one has to go back to the beginning of the evolution of the horse family. Enormous changes took place regarding size and structure of the animal in response to environmental stresses, such as changes in climate and food sources. We know much about the nutrition of the ancestors of the horse thanks to well-preserved fossil remains which were found in the river valley clays, sand and sandstones of the rich palaeontologic beds of the old and new worlds, particularly in North America.

The oldest known fossils of the ancestral horse date from the Eocene period, about 60 million years ago. This species is referred to as *Eohippus* (the 'dawn horse') or *Hyracotherium*. *Eohippus* differed from the modern horse in three important respects:

- it was much smaller;
- about the same size as a present day fox;
- it had well-developed digits, with four hoofed toes on each fore foot and three on each hind foot;
- it had a different dentition, in particular the cheek teeth lacked the hard serrated grinding surfaces typical of the modern horse.

From historical records it has been deduced that although the preferred diet of *Eohippus* was leaves and soft vegetation, it was also able to digest fruits and seeds; its non-grinding molars would have been perfectly adequate for this.

With time the climate changed the tropical forests into grass and bush steppe. These environments had few trees and mainly supported the growth of tough wiry grasses. As *Eohippus* evolved in response to these new conditions, certain changes gradually occurred:

- the toes joined to form the familiar single-toed hoof which allowed the animal a greater turn of speed to evade predators;
- there was a steady enlargement of the body;
- the molar teeth became longer, wider and more ridged, to enable the animal to grind the tougher vegetable fibres found in the grass and scrub that it was now forced to eat.

Supported by the gradual changes in its food spectrum the differentiated structure of the digestive tract of the horse was now able to cope with different feeds such as concentrated substances and fibrous plants which were digested with the help of enzymes. These feeds varied during the different seasons of a year.

Compared with the digestive tract of ruminants, horses have problems digesting products of the grass and bush steppe with high crude fibre content, low-grade protein and low levels of carbohydrates, starch and fat. Equids are able to reduce these disadvantages by selectively grazing large amounts of feed each day.

Little is known either about the early development of specific strains of horses or the first domestication by man. Undoubtedly, the earliest association between man and horse was one-sided, that of hunter and prey. This is well illustrated by cave paintings such as those in France and Spain, which are thought to date from 15,000 BC or even earlier. Vast repositories of horse bones have been found near these cave shelters, the most notable being at Solutre in France, where the bones of over 40,000 horses were discovered.

The initial domestication of the horse probably occured in several parts of the world at the same time. Certainly two of these areas were China and Mesopotamia between 4500 and 2500 B.C., where horses were mainly used to carry goods. The domestication of the horse was later than other species (such as the dog) probably as a result of its large size and somewhat unpredictable temperament.

Around 1000 B.C. man started to use horses in a more intensive way, e.g. pulling and riding. This intensive use of horses led to a change in the spectrum of feed selection and feeding methods (Meyer, 1992), because:

- roughage and grass were no longer able to fulfil the energy requirements of the working horse;
- the capacity for intake of roughage and grass was limited through the physiology of the horse which, in turn, meant less time for work;
- the intake of feed in large volumes reduced the efficiency of the animals.

The feeding of concentrated feed—in addition to roughage—started around 2000 BC. For several centuries feedstuffs such as wheat, oats, barley, peas, dates and millet were given to the horses.

In central Europe oats were the dominant foodstuff because of favourable cultivation conditions and good nutritional properties. This system of feeding has survived virtually unchanged to the present day. It is essentially a three-component regime consisting of hay or grass (fibre or 'roughage'), cereals (or 'concentrates') such as oats, barley and wheat, to provide further energy and protein, and a mineral/vitamin supplement (Table 8.1). The latter is designed to complete the diet by rectifying any deficiencies in the other components. This is particularly important if the horse is under exacting conditions (such as hard work, breeding or growth) or if the other components (especially the hay or grass) are of poor or suspect quality.

TYPICAL RATION FOR THE ADULT HORSE, BASED ON AMOUNT OF ACTIVITY			
Activity	**Roughage (e.g. hay) (kg)**	**Concentrates (e.g. cereals) (kg)**	**Supplements (g)**
Maintenance	1.8–2.5	–	60
Light work	1.0–1.5	0.5–0.75	60
Moderate work	1.0	1.0	120
Hard work	1.0	>1.25	300

TABLE 8.1: Typical ration for the adult horse. Amounts per 100 kg body weight per day.

Like dogs, horses are unusual among mammalian species in that they have a wide range of bodyweights. Adult miniature horses can weigh less than 50 kg, while the bodyweight of mature draft animals such as Percheron and Shire horses can be over 1000 kg. The lifespan of the horse is generally greater than that of dogs or cats and is typically 20–30 years.

Anatomy and Functions of the Digestive Tract

Horses are non-ruminant herbivores, as are rabbits and guinea pigs; this means that they do not have a rumen, unlike sheep and cattle, with a bacterial population to help utilise plant fibres. However, horses have a relatively large caecum (and colon) containing the necessary bacterial population for the digestion of plant material. In contrast, dogs and cats have a relatively simple digestive tract and cannot effectively digest large amounts of roughage. The main function of any digestive tract is the mastication, digestion, absorption and inital storage of the nutrients. The following sections describe the individual components of the horses alimentary canal, with respect to these essential functions (see also Chapter 3).

Mouth

Food is grasped with the help of lips, tongue and teeth. Horses have lips which are relatively strong, mobile and sensitive and they can therefore be selective in choosing feed. The greatest advantage of this is a reduction of the risk of consuming foreign bodies; the disadvantage is a selective grazing which reduces the grass-species variety on the pasture.

Horses masticate their feed with the cheek teeth and pulverise it into small particles. The masticated feed is wetted by saliva produced by three pairs of glands. The quantity of saliva is quite large (ponies may secrete as much as 12 l per day). Horses' saliva contains the enzyme amylase, but in such a low concentration that its role in the process of the digestion of carbohydrates is very limited.

Saliva mainly contains minerals and bicarbonates, which probably play an important role in buffering the amino acids in the first part of the stomach. The process of mastication in horses is important: if too short, there is a risk of abnormal behaviour such as licking and nibbling the stall—so-called cribbiting—often wind-sucking accompanies this and can interfere with normal digestion.

Oesophagus

The oesophagus in the mature horse is relatively short, approximately 1.5 m. It is composed of strong ring muscles which are responsible for the transport of the feed particles into the stomach.

At the junction between the oesophagus and stomach there is a strong sphincter. It is almost impossible for stomach contents or gas to be forced from the stomach into the oesophagus. Therefore the vomiting reflex is very uncommon in horses and distension of the stomch can be so severe that it will rupture before vomiting occurs.

Stomach

The stomach of the adult horse is relatively small (about 15 l) and provides about 8% of the capacity of the gastro-intestinal tract. Although the horse's stomach is seldom completely empty, emptying of the stomach starts soon after feeding. The movement of a meal through the stomach and small intestine into the large intestine is rapid, which results in only limited contact with the gastric secretions.

It is generally recommended, therefore, that horses are fed several meals a day because their digestive system is adapted to a frequent intake of small amounts of feed. For example, Meyer (1992) recommends not to feed more than 0.4–0.5 kg/100 kg BW feed per meal.

Concentrated feeds must be of good hygienic quality, otherwise the activity of the intestinal microbes can be adversely affected, leading to digestive upsets typified by colic.

Small intestine

The small intestine, which is about 20 m long in a mature horse, provides nearly 30% of the capacity of the gastro-intestinal tract. It is divided into the duodenum, jejunum and ileum; bile and pancreatic ducts open into the duodenum.

Bile salts, which are necessary to emulsify lipids, are secreted by the liver and, as horses do not have a gall bladder, are continuously released into the digestive tract. The rate of bile secretion is about 300 ml/hr. The pancreas produces a juice which contains the enzymes trypsin, lipases and amylases, which are responsible for the digestion of protein, fat and carbohydrates. Alkali and bicarbonates, which buffer the acid ingesta, are also found in the pancreatic juice and lead to the preservation of an optimal environment for the functioning of digestive enzymes (see Chapter 3). Digestive enzymes are also secreted by intestinal cells, but the pancreas releases by far the greatest amount—up to 10% of the bodyweight per day. The contents of the small intestine are quite fluid (4–8% dry matter), depending on the ration.

Large intestine

The large intestine, about 7 m long, is divided into the caecum, right ventral colon, left ventral colon, right dorsal colon, transverse colon, small colon and rectum. The caecum is a large pouch (capacity 30 l) at the junction of the small intestine and colon. The colon has the greatest capacity of the digestive tract. The dry matter content of the ingesta increases as the material travels from the caecum to the rectum. Water absorption mainly takes place in the last part of the large intestine, in the small colon and rectum. The ingesta of the caecum usually contains 6–10% dry matter. The caecum and the largest part of the colon are fermentation chambers where, with the help of bacteria and protozoa, mainly structured feed ingredients, e.g. cellulose, hemicellulose, pectins are digested. The microbial population of the caecum and the rumen are qualitatively similar. Larger amounts of water-soluble vitamins can be synthesised by bacteria. The vitamin B content of horse faeces is therefore much greater than the vitamin B content of feed.

The efficiency of the utilisation is relatively unclear, but normally no vitamin B deficiencies occur during normal functioning of the large bowel (see vitamin section).

Rate of passage

The overall rate of passage through the intestine is about 35–50 hr. Approximately 85% of this time is in the large intestine (Meyer, 1992).

Digestibility of Feedstuffs

One important basis for the evaluation of energy and nutrient requirements is a good knowlegde of the digestibility of different nutrients in different states and feeding regimes. In the nutrition of wild horses starch plays a subordinate role whereas in performance horses up to 40% of the overall energy requirement is provided by starch (Zmija, 1991). Some variations of the colic-syndrome and most *diet-related* laminitis cases can be explained by overloading the capacity for starch digestion of the small intestine. Laminitis is an inflammation or oedema of the sensitive laminae (internal tissue layers) of the hoof which are susceptible to a metabolic toxaemia

EFFECT OF SOURCE AND PREPARATION ON PRE-ILEAL STARCH DIGESTIBILITY IN THE HORSE		
Source	**Preparation**	**Digestibility (%)**
Oats	Whole	83.5 (9.5)
	Rolled	85.2 (13.4)
	Ground	98.1 (1.6)*
Maize	Whole	28.9 (25.0)
	Rolled	29.9 (19.8)
	Ground	70.6 (23.2)*

TABLE 8.2: The effect of source and preparation on pre-ileal starch digestibilty. Values stated as mean (SD). Significantly different from other treatments, $p < 0.05$. (From Kienzle *et al.*, 1992.)

caused by the aforementioned digestive problems. It is a major health problem in the horse as it seriously affects the overall well-being and performance of the animal.

A high precaecal digestibility improves the utilisation of energy and reduces the risk of microbiological disorders in the large bowel. Preileal digestibility of oat starch is significantly higher than maize starch (Kienzle *et al.*, 1992). A considerable improvement of preileal starch digestibility is achieved in both cereals by thorough grinding while rolling or breaking had little effect (Table 8.2).

Protein is broken down by enzymatic digestion in the small intestine and fermentative digestion in the large intestine. The end result of enzymatic protein digestion in the small intestine is the absorption of a spectrum of amino acids that reflects the amino acid profile. The end result of the fermentative protein digestion in the large intestine is the production of ammonia; hence no effective absorption of essential amino acids takes place. The digestibility of different proteins passing through the digestive tract can be very similar; but biological values can differ markedly.

As a result of different experiments with horses fed mixed diets it is estimated that the true digestibility of nitrogen in the small intestine ranges from 45–80%; in the large intestine from 80–100%. Fifty to seventy percent of the digestible protein was digested in the small intestine (Potter *et al.*, 1992).

Horses seem to digest and utilise supplemental amounts of fat or oil effectively. The digestion of fat is influenced by the source and quality of the fat, the amounts fed and the composition of the diet. Dietary lipids are mainly digested and absorbed in the small intestine. The composition of body fat is influenced by the composition of dietary fat because the fatty acids are absorbed from the small intestine before they can be altered by the bacteria of the large bowel.

The digestibility of nutrients can be influenced by several factors. High fibre content in a feedstuff reduces digestibility (e.g. straw is only 35%). Physical treatment can improve the digestiblity of small grains (see Table 8.2) which can otherwise be consumed without being crushed by the teeth. Pelleting of hay decreases fibre digestion because the rate of passage will be increased. High intake levels may decrease digestibility; the rate of passage is increased and the bacteria do not have as much time to act on the fibre.

Energy Requirements

The energy content of foods for humans is based on classical Atwater energy conversion factors of 4 kcal (16.7 kJ) per gram of food for protein, 3.75 (15.7 kJ) for carbohydrate and 9 (37.7 kJ) for fat; these factors are based on the assumption that foods consumed by humans are highly digestible. In the horse most feedstuffs are not highly digestible. Therefore differences in digestibility must be taken into account. Digestible energy (DE) values are normally used in feed for horses.

Maintenance

Horses are considered to be on a maintenance energy requirement when they are under no particular extra demands of lifestyle or lifestage i.e. they are not performance horses, they are not growing or lactating or in the latest stage of pregnancy.

Under practical conditions, horses need, on average, 0.55–0.63 MJ DE per kg BW $^{0.75}$ in an environmental temperature inbetween $-10°C$ and $+25°C$ (Meyer, 1992).

Muscle work

Chemical energy is transformed in the muscle

into mechanical energy. In the muscle energy is mainly delivered by adenosine triphosphate (ATP) which is transformed into adenosine diphosphate (ADP) and immediately regenerated via creatine phosphate (CP) to adenosine triphosphate. Further energy for the synthesis of ATP and CP is delivered by glycogen which is stored in the muscle and metabolised into glucose. Glycogen can be metabolised aerobically (1 mol glucose = 38 mol ATP) or anerobically (1 mol glucose = 2 mol ATP). Both anaerobic and aerobic work capacities can be enhanced by training. The estimation of DE requirement for work is complicated, because it includes many factors that are difficult to quantify (e.g. condition, training, rider weight, environmental temperature, etc.). The DE requirements for work can be estimated by several methods. It is useful to divide work into light, medium and hard; the differences in feed required are shown in Table 8.1.

Nevertheless, a good guideline for the appropriateness of the energy ration is the weight maintenance and general condition of the horse.

Carbohydrates

Horses use dietary carbohydrates as a primary energy source. As a result of the digestion of simple dietary carbohydrates in the small intestine, glucose enters the portal vein and is a significant contribution to the horse's energy budget.

Some simple dietary carbohydrates, which are not digested precaecally, and complex carbohydrates, such as cellulose, hemicellulose and pectins, are digested mainly in the caecum and colon by the process of anaerobic microbial fermentation, which results in the production of volatile fatty acids: principally acetic, propionic and butyric acids. These compounds are absorbed and constitute important sources of energy, particularly for horses fed high-forage diets. Of the volatile fatty acids listed above, only propionic can be converted into glucose, the others are used as energy sources via aerobic oxidation.

Foals are able to digest lactose, but as in dogs and cats, this ability declines with age.

Horses older than 3 years have a limited lactase activity (Roberts 1975). Sudden introduction of lactose-containing feedstuffs— more than 2 g lactose/kg BW—to mature horses may induce digestive disturbances (Meyer,1992).

Fat

Dietary lipids can be an important alternative source of energy in horses.

Fat can reduce the protein:energy ratio and permits a reduction in the starch content of the diet by increasing the energy density of the feedstuff. It has been noted that restricting horses to 15% less feed to maintain bodyweight when fed a diet containing 8% feedgrade animal fat (Hintz *et al.*, 1978), reduces the bulk in the large bowel, and may therefore lessen the risk of colic.

During the last few years, several investigations have been made to answer the question of how dietary lipids can influence performance i.e. ability for sustained exercise in horses. Higher muscle glycogen concentrations in horses before exercise were reported by Hambleton *et al.* (1980), when 12% fat was added to the diet. Griewe *et al.* (1989) studied the effects of fat in the diet of 2-year-old horses beginning training and observed muscle glycogen concentrations during the final exercise, test decreased only by 19% in the fat-fed horses as compared with 65% in the control group. Increased muscle glycogen stores, which enhanced the anaerobic performance potential, were observed by Oldham *et al.* (1990) while adapting racehorses to a fat-supplemented diet. Feeding two different sources of fat (soybean oil and coconut oil) alone and as a mixture, was found to affect the metabolic response to a standard exercise test in comparison with a high carbohydrate diet in thoroughbred racehorses. During a five-day standardised exercise test the speed was significantly lower in the control group than for all three fat treatments (Pagan *et al.*, 1993).

Most of the studies evaluating the feeding of fat to horses have shown some advantages but have also been of short duration. Long-term studies are required to investigate the extended effects of feeding high fat diets to performance horses and to define the upper fat limits in diets.

Lawrence *et al.* (1989) fed a control or fat-supplemented diet to growing horses. No differences in feed intake, feed efficiency or average daily weight gain were observed. However, feeding fat at the expense of carbohydrates stimulated early growth in yearling horses. The effects were not maintained as the horses matured (Scott *et al.*, 1989).

The influence of addition of fat on reproductive performance during late gestation and early lactation was investigated by Davison *et al.* (1991). Mares fed the fat-supplemented diet consumed less feed during the gestation period, compared to mares fed the control diet while maintaining body fat content. The energy intake during early lactation for the fat-fed group was greater, but no gain in bodyweight or body fat occured. However, the milk fat content was increased in this group. There were no dietary effects on birth weight of foals or weight gains to day 60. The mares of the fat-supplemented diet showed a tendency to require fewer cycles per conception (1 vs. 2) and to have a higher pregnancy rate (100 vs. 89%).

The precise essential fatty acid requirements of the horse are not known; a dietary linoleic acid content of at least 0.5% in dry matter is recommended (NRC, 1989).

Protein

The dietary protein requirement of a horses is a function of the needs of the animal, the quality of protein available, and the digestibility of that protein (NRC, 1989). Protein that is not used in building body tissues, can be used to supply energy. The horse is able to manufacture some amino acids (non-essential). Other amino acids (lysine, methionine, tryptophan, leucine, isoleucine, threonine, valine, histidine and phenylalanine) are essential and must be supplied in the feed. The primary site of protein digestion, which starts in the stomach, is the small intestine. The amino acids that are absorbed represent the amino acids in the diet and are not altered by

bacteria. However, there also appears to be some absorption of amino acids of bacterial origin from the large intestine of the horse. Nevertheless, some studies have reported that the bacterial amino acids synthesised in the large intestine are not effectively utilised by the horse (Wysocki and Baker, 1975).

Horses with elevated protein requirements —e.g. foals and broodmares—are particularly dependent on a supply of essential amino acids. Young foals can absorb intact protein up to 36 hr after birth. It is important that the foal receives colostrum as soon as possible after birth. This substance contains maternal antibodies (immunoglobulins) which provide protection against infection for young animals. There is no transmission of antibodies to the foal across the placenta.

The protein requirement for maintenance is determined by no net gains or losses in body nitrogen, and excluding any protein which is contained in the milk. The animal must replace shed epithelial cells and hair, must provide for various secretions and keep all the cellular tissues in a state of dynamic equilibrium. Horses during maintenance need about 5 g digestible dietary protein per MJ digestible energy. This is much lower than the requirements for dogs or cats.

Protein deficiency is rare and becomes clinically relevant only if it reaches extremes. Under practical conditions it is more likely that too *much* protein is fed to mature horses. Under maintenance conditions healthy horses can tolerate up to 3 times the recommended levels. Exercise has little influence on protein requirements. An apparent increased nitrogen retention in working horses was documented by Freeman *et al.* (1988). This retention included muscle hypertrophy as animals became physically conditioned, perhaps an increased muscle protein content and some nitrogen loss via sweat.

Sweat contains about 1–1.5 g of nitrogen/ kg and losses of up to 5 kg sweat/100 kg BW have been estimated (Meyer, 1987).

The bacteria in the large bowel need a specific dietary digestible protein (DP):DE ratio for optimal activity. Meyer (1992) recommends a protein/energy ratio of 5 g per

MJ, i.e. the same as the maintenance requirement.

Requirements for reproduction and growth

Pregnancy (gestation) in horses lasts about 11 months. For pregnant broodmares, maintenance energy levels are adequate until the last 90 days of gestation, when most of the foetal tissue (60–65%) develops. The energy requirements are elevated up to 1.5 times that of maintenance during this time and the DP:DE ratio should be 6–7 g per MJ.

The protein content of milk is highest immediately after parturition (birth) and decreases gradually as lactation progresses. On the other hand, the milk production increases so that the overall requirements for protein remain constant for the first 2 months of lactation. Also more energy is required for the first period of lactation. The level increases slightly during the first 3 months then steadily decreases. To fulfil the requirements of essential amino acids it is recommended to feed protein sources of high quality during lactation.

The major nutrient factors influencing the growth of young horses are protein and energy intakes. The desired rate of weight gain influences the feeding regime. Normally, height increases faster than bodyweight and therefore proper nutrition during the period of rapid bone elongation is essential for developing a sound skeleton; an excessively rapid weight gain may be responsible for an increased incidence of skeletal problems in young foals. For example, at one year of age foals reach about 90% of the height of adult horses whereas their bodyweight is only about 60–75% of the final adult weight.

The daily protein requirements are highest during the first month (DP:DE 10 g per MJ) and then decrease slightly (6 g per MJ). An adequate supply of essential amino acids is particularly crucial for the first 8 months of life.

Major minerals and trace elements

Because the skeleton is of such fundamental importance for the performance of the horse, its mineral requirements deserve very careful attention.

Excessive intakes of certain minerals may be as harmful as deficiencies—therefore mineral supplements should be based on the basic feeds in the diet. The total mineral contribution and availability from all aspects of a diet is the basis on which to evaluate the mineral intake of the horse.

Calcium and phosphorus

Calcium and phosphorus are absorbed from the small intestine. Calcium is actively absorbed with the help of a calcium-binding protein produced by mucosal cells. The interrelationship of these two minerals is important for the mineralisation of bones. Calcium, for example, makes up about 35% of the bone structure. Both minerals are also involved in other important body functions as muscle contractions, nerve functions and blood clotting mechanisms.

The need during early growth is therefore much greater than for maintenance of the mature animal. The maintenance requirements for calcium and phosphorus depend on balancing losses in the faeces and urine. Working horses lose some calcium and phosphorus through sweat, however the obligatory increase in calcium intake in dry matter consumption to meet energy requirements appears to be enough to fulfil the extra requirements.

Special calcium requirements occur in late gestation, for mineralisation of the foetal skeleton, and during lactation. The calcium and phosphorus requirements are also higher for the first half-year of life. The calcium:phosphorus ratio should be maintained at not less than 1:1 and can be tolerated up to 3:1. Care should be exercised when evaluating phosphorus levels in feed because of phytate forms of phosphorus, which are relatively unavailable to the horse.

Magnesium

About 60% of magnesium found in the body

mass is associated with the skeleton. The true absorption rate from feeds ranges from 40 to 60%. Supplementary dietary magnesium can be absorbed at approximately 70%. The daily magnesium requirement for maintenance is about 15 mg/kg of bodyweight. The magnesium requirements for mares during the final phase of pregnancy are slightly increased, and for lactation and growth larger amounts of dietary magnesium are required.

Sodium and chloride

The sodium chloride concentration in natural feedstuffs for horses is often lower than 1%. The daily sodium requirement for maintenance is estimated to be about 20 mg/kg bodyweight/day. For chloride, Meyer (1992) recommends 80 mg/kg bodyweight/day to avoid disturbances in the acid base balance. Endogenous sodium loss has been estimated at 15–20 mg/kg bodyweight/day. Prolonged exercise and elevated temperatures dramatically increase the requirements for sodium and chloride. A hard-working 500 kg horse can lose up to 50 g sodium and 75 g chloride in a few hours. For these horses an extra salt lick is recommended, because other feedstuffs are not able to fulfil these requirements. The sodium requirement is only slightly increased for foals and broodmares.

Potassium

Forages contain 1–2% potassium, whereas cereal grains contain 0.3–0.4%. The daily potassium requirements for maintenance are estimated to be about 50 mg/kg bodyweight. For gestation these values are only slightly increased, whereas growing horses and lactating mares require higher potassium levels. The requirements for hard-working horses can increase to more than twice maintenance levels.

Because forages usually consitute a significant portion of the diet, the potassium requirements for horses should normally be met by a standard regimen.

Iron

About 60% of the iron content of a horse is distributed in haemoglobin, 20% in myoglobin, 20% in storage and transport forms and 0.2 % in cytochromic or other enzymes. A deficiency of iron causes anaemia and because it is a major constituent of an oxygen carrier, this deficiency is especially relevant for growing and race horses. Forage contains up to 250 mg of iron/kg, grains usually contain less than 100 mg. As a dietary concentration of 50 mg/kg is adequate for all stages of the lifecycle (NRC 1989), common feedstuffs should normally meet the iron requirements.

Copper

Copper has many functions in the body. It is essential for several enzymes which are important for bone, cartilage and elastin formation, utilisation of iron and the formation of the pigments of the hair and melanin.

The copper concentration of common feedstuffs ranges widely (4 mg/kg for corn; 80 mg/kg for cane molasses). Copper metabolism can interact with many other minerals including iron, molybdenum, sulphate, zinc and selenium (see Chapter 2).

Requirements for copper are estimated to be 10 mg/kg diet for foals and broodmares and 5 mg/kg diet for all other horses.

Zinc

Zinc is required for several enzyme systems. High concentrations of zinc occur in epidermal tissues (skin and hair, bone, muscle, blood and internal organs). Deficiencies can cause parakeratosisis, characterised by lesions of the superficial layers of the epidermis, hair loss and infections. Common equine feedstuffs contain 15–40 mg zinc/kg dry matter and, on average, zinc requirements are around 50 mg/kg dry matter, so some supplementation is usually required.

Manganese

Manganese is essential for carbohydrate and lipid metabolism and the formation of cartilage. Roughage contains 40–140 mg of manganese/kg dry matter, oats 50 mg/kg dry matter. As 40 mg per kg/dry matter is considered to be sufficient to fulfil nutritional requirements, under normal circumstances no deficiencies should occur.

Cobalt

Cobalt is an integral part of vitamin B_{12} and microorganisms in the digestive tract use cobalt in the synthesis of this vitamin. Cobalt requirements are estimated at 0.1 mg/kg dry matter of feed. Normally common feedstuffs should meet the cobalt requirements.

Iodine

Iodine is a component of thyroxine (T_4) and triiodothyronine (T_3). These thyroid hormones are essential for reproduction and normal physiological processes. The iodine concentrations of forage range from 0 to 2 mg/dry matter, depending on the iodine content of the soil on which the feed was grown. Iodine requirement for horses is estimated at 0.1–0.6 mg/kg dry matter, although these values are based largely on data from other species (NRC 1989).

Excessive levels of iodine are particularly toxic to horses and a maximum content of 5 mg/kg diet is recommended.

Selenium

Selenium forms an integral part of the glutathione peroxidase molecule which aids in detoxification of peroxides.

The selenium requirement of the horse is estimated to be about 0.1 mg/kg dry matter. Disturbances can be expected at values less 0.05 mg/kg. The concentrations of selenium in feedstuffs vary a lot and are influenced by soil selenium and pH. The range found varies from below to several times the requirement so it is important to assess the level to ensure an adequate supply. This is further emphasised by the high toxicity of selenium. In common with other species this trace mineral is dangerous to the horse at relatively low intakes, and a maximum dietary content of

2 mg/kg dry matter is recommended (NRC, 1989).

Fat-soluble vitamins

The absorption of fat-soluble vitamins can be affected if there are diseases or metabolic abnormalities that interfere with fat absorption.

Vitamin A

Vitamin A is required to maintain cellular structures and is therefore important for vision, cell differentiation and fertility. In a vitamin A deficiency, epithelial cells may develop as keratinised squamous cells.

Plants do not contain vitamin A, but they contain the yellow pigment carotene, which the horse can convert to vitamin A. All beta-carotenes are formed in highest concentrations in green forage. Carotene is easily oxidised or destroyed. As much as 80% of the carotene in hay can be lost during harvesting and the carotene content of hay stored for six months may be reduced by half. Corn contains some carotene; the other grains contain little or none. The conversion of carotene to vitamin A in the horse is not totally efficient and the National Research Council has estimated that 1 mg of carotene is equivalent to 400 IU of vitamin A (NRC, 1989).

Vitamin A requirements are especially high in broodmares and foals. Infection also increases the vitamin A requirement. Pregnant and lactating mares should be supplemented with vitamin A and beta-carotene in winter, because intakes from feed are likely to be lower at this time of year. The importance of adequate supplementation is heightened by the fact that the vitamin A supply of suckling foals depends directly on the vitamin A content of the mare's milk.

There is some evidence that beta-carotene itself may have some influence on fertility in the horse regardless of the vitamin A status of the animal, but this has yet to be confirmed.

As in other species, excessive intakes of

vitamin A are toxic to horses with adverse effects on bone structure, hair and skin condition, and muscle tone. The NRC (1989) has proposed a maximum vitamin A content of 16000 IU per kg dry matter of feed. It is theoretically possible to exceed the carotene equivalent of this value with some types of pasture, e.g. alfalfa. However, no carotene toxicity has been linked to this source and this is probably due to regulation of carotene conversion to vitamin A in the horse's intestine.

Vitamin D

There are two sources of this vitamin available to the horse. Vitamin D_2 (ergocalciferol) is formed by the action of ultraviolet light on the sterol ergosterol found in plants. It is found in high concentrations after plants have been cut and exposed to sunlight and in the lower, dead leaves of living plants. The irradiation is ineffective in living plants and it is assumed that chlorophyll screens out the active part of the spectrum. Vitamin D_3 (cholecalciferol) results primarily from ultra-violet irradiation of 7-dehydrocholesterol synthesised by the tissues of the horse and present in the skin.

Both vitamins are hydroxylated in the liver and the resulting active components are required for the formation of a calcium-binding protein which aids the absorption of calcium and phosphorus. A deficiency can, for example, result in poorly mineralised bones and swollen joints. In contrast, excessive intake of vitamin D is characterised by calcification of the blood vessels, heart and other soft tissues, and by bone abnormalities.

Vitamin D requirements of horses are not clearly established. Horses kept outside (i.e. exposed to sunlight) or on forage or pasture containing high vitamin D_2 levels are unlikely to be deficient. Requirements of growing foals and pregnant or lactating broodmares are higher than other lifestages and dietary levels of 600–800 IU/1kg dry matter are recommended. In view of the toxicity of high intakes of this vitamin, a maximum safe level of 2200 IU of vitamin D_3/kg of diet (dry matter) has been proposed for long-term feeding (NRC, 1989).

Vitamin E

The vitamin E activity of feeds and feedstuffs depends on the conditions of storage and the chemical forms present. Naturally-occurring forms tend to be somewhat unstable, especially if there is an increased likelihood of oxidation, for example if the structure of the grains has been disrupted, permitting greater peroxidation of unsaturated fats. Horses have a requirement for vitamin E, but the precise value is not known. Requirements seem to increase during stress, exercise and growth, pregnancy and lactation. The minimum requirement of 50 IU per kg diet dry matter may not be sufficient for optimum health under all conditions and a safer value is probably nearer 100 IU per kg.

Vitamin E deficiency in horses produces disturbances in muscle action and some effects on the nervous system: abnormal gait and uncoordinated movement of the limbs are typical signs. Vitamin E is much less toxic than the other fat-soluble vitamins and adverse effects have not been reported in the horse. Nevertheless, in view of effects in other species a maximum dietary content of 1000 IU per kg diet is recommended (NRC, 1989).

Vitamin K

Dietary vitamin K requirements for horses have not been determined because, under normal circumstances, the synthetic rate of vitamin K in the large bowel is sufficient to fulfil the requirements of horses. Also pasture and good quality hay contain high concentrations of vitamin K.

Water-soluble vitamins

Normal intestinal synthesis and the quantities present in typical horse feeds are normally adequate to fulfil the B-vitamin requirements of horses.

Extra supplementation can be needed for

young foals, hard-working horses and horses with diarrhoea.

Vitamin B₁ (thiamin)

Vitamin B_1 plays a central role in the metabolism of carbohydrates. About 25% of the free thiamin synthesised in the caecum is absorbed into the blood, but this may not invariably supply the total requirement and a dietary need has been established. Horses fed with poor quality hay can develop signs of thiamin deficiency. This results in loss of appetite and weight, loss of co-ordination of the hind legs, a dilated and hypertrophied heart and a decline of the activity of the enzymes that require thiamin as a cofactor.

A total dietary level of 3–5 mg/kg seems to meet the requirement. The higher value may be more appropriate for performance and working horses.

Other B complex vitamins

There are no reliable dietary recommendations for the other B vitamins in the horse. Deficiency signs have not been observed and it is assumed that bacterial synthesis in the intestine satisfies the animal's requirements under normal conditions. It is only if this synthesis is compromised in some way, for example by disease or drug therapy, that a dietary supply may be necessary. In this way the horse is very different from the other species discussed in this book and demonstrates its 'quasi-ruminant' metabolism.

One interesting exception is the effect on hoof quality by large intakes of biotin (up to 30 mg per day). Structural integrity has been reported to be improved by long-term biotin treatment in adult horses.

Ascorbic acid

Horses have tissue requirements for ascorbic acid, but like dogs and cats, can synthesise vitamin C from glucose. Suckling foals get large amounts of vitamin C via colostrum. It has been suggested that vitamin C supplementation may be of some value to horses under certain conditions, for example when subject to infection or displaying poor performance, in high environmental temperatures, or when training. However, the subject remains controversial and no dietary requirement has been conclusively identified.

Water

The body water content of the horse is typically about 70% of the total weight on a fat-free basis. As with other terrestrial mammals, an adequate supply of clean water is essential. The horse also needs to take water with its food to act as a fluid medium for digestion and for the transport of digesta through the gastro-intestinal tract. Water is also essential for temperature regulation (sweat production). As a general guide, horses need about 2–3 l of water per kg dry diet each day but various factors such as larger amounts of roughage in the diet can increase this figure by up to 25%.

Main losses are via the kidneys, intestine, skin and lungs. Normally 40–50% of the water intake consumed is excreted as urine. Faeces normally contain 60 to 80% water; 5–25 ml of water per kg bodyweight per day are excreted via this route.

The main factors influencing water requirement (other than disease) are lactation, environmental temperature and exercise. Lactating mares lose up to 20 kg per day via their milk. Water losses of up to 10% of the the bodyweight have been found in horses during endurance rides and these represent increases of up to three times the values at rest. Similar increases have been observed in horses kept in a high temperature environment (38°C). These considerable changes emphasise the potential scale of water turnover in the horse and endorse the importance of adequate supplies to the husbandry and activity of the animal.

Concluding remarks

It is rather fitting that this book should conclude with a discussion on horse nutrition as, in some ways this species represents an interesting link between companion and domestic animals. The main aim of pet nutrition is a long and healthy life; in farm animal nutrition, growth and productive performance are the principal criteria, although it necessarily includes health and welfare. In the horse, we have a 'hybrid' which frequently incorporates both the companion and performance elements of feeding and nutrition. Clearly, the pet aspects remain strong, perhaps more so than in the past. On the other side of the equation, the practical feeding of horses is still influenced by factors such as pasture quality, the presence of dietary fibre ('roughage') in the ration and physical characteristics of the feed, which are not normally encountered for other pets. In particular, it is the association with performance that is a major area for future research in horse nutrition. In view of the dual objectives of feeding horses, it is important that this research should focus on the well-being of the animal just as much as its athletic potential.

Bibliography

AAFCO (1993) Official Publication Association of America Feed Control Officials Inc.: Atlanta, U.S.A.

Altman, R. B. (1978) Non infectious diseases (perching birds). In *Zoo and Wild Animal Medicine*. ed. M. E. Fowler. 1st Edn. pp. 384–388. Philadelphia: W. B. Saunders.

Anderson, P. A., Baker, D. H., Sherry, P. A. & Corbin, J. E. (1979) Choline–methionine interrelationship in feline nutrition. *Journal of Animal Science*, **49**, 522–527.

Aschoff, J. & Pohl, H. (1970) Rhythmic variation in energy metabolism. *Federation Proceedings*, **29**, 1541–1552.

Baines, F. M. (1981) Milk substitutes and the hand rearing of orphaned puppies and kittens. *Journal of Small Animal Practice*, **22**, 555–578.

Barry, R. E. (1976) Mucosal surface areas and villous morphology of the small intestine of small mammals: functional interpretations. *Journal of Mammalogy*, **57**, 273–290.

Batt, R. M. & Horadagoda, N. U. (1986) Role of gastric and pancreatic intrinsic factors in the physiologic absorption of Cobalamin in the dog. *Gastroenterology*, **94**, A1339.

Blackmore, D. K. (1963) The incidence and aetiology of thyroid dysplasia in budgerigars (*Melopsittacus undulatus*). *Veterinary Record*, **75**, 1068–1072.

Blaza, S. E., Booles, D. & Burger, I. H. (1989) Is carbohydrate essential for pregnancy and lactation in dogs? In *Nutrition of the Dog and Cat*. eds. I. H. Burger & J. P. W. Rivers. pp. 229–242. Cambridge: Cambridge University Press.

Brooke, M. & Birkhead, T. (1991) *The Cambridge Encyclopaedia of Ornithology* Cambridge: Cambridge University Press.

Buddington, R. K., Chen, J. W. & Diamond, J. M. (1991) Dietary regulation of intestinal brush-border sugar and amino acid transport in carnivores. *American Journal of Physiology Regulation and Integrated Comparative Physiology*, **261**(4), R793–R801.

Buffington, C. A., Branam, J. E. & Dunn, G. C. (1989) Lack of effect of age on digestibility of protein, fat and dry matter in Beagle dogs. In *Nutrition of the Dog and Cat*. eds. I. H. Burger & J. P. W. Rivers, p. 397. Cambridge: Cambridge University Press.

Buffington, C. A., DiBartola, S. P. & Chew, D. J. (1991) Effect of low potassium non-purified diet on renal function of adult cats. *Journal of Nutrition*, **121**, S91–92.

Burger, I. H. & Barnett, K. C. (1982) The taurine requirement of the adult cat. *Journal of Small Animal Practice*, **23**, 533–537.

Burger, I. H. & Thompson, A. (1993) Reading a pet food label. In the Waltham Book of Clinical Nutrition. eds. J. M. Wills and K. Simpson. Oxford: Pergamon.

Burger, I. H., Blaza, S. E., Kendall, P. T. & Smith, P. M. (1984) The protein requirement of adult cats for maintenance. *Feline Practice*, **14**, (2) 8–14.

Burger, I. H. & Johnson, J. J. (1991) Dogs large and small: the allometry of energy requirements within a single species. *Journal of Nutrition,* **121,** S18–S21.

Burger , I. H. & Rivers, J. P. W. (eds.) (1989) *Nutrition of the Dog and Cat.* Waltham Symposium 7. Cambridge: Cambridge University Press.

Burger, I. H. & Smith, P. M. (1990) Amino acid requirements of adult cats. In *Nutrition, Malnutrition and Dietetics in the Dog and Cat.* eds. H. Meyer, E. Kienzle, & A. T. B. Edney. pp. 49–51. London: British Veterinary Association.

Buttery, P. J. & Boorman, K. N. (1976) The energetic efficiency of amino acid metabolism. In *Protein Metabolism and Nutrition.* eds. D. J. A. Cole, K. N. Boorman, P. J. Buttery, D. Lewis, R. J. Neale, & H. Swan. pp. 197–206. London: Butterworths.

Buttner, E. E. (1968) Arginine deficiency link with feather plucking. *Cage and Aviary Birds,* **134,** 256.

Calles-Escandon, J. & Horton, E. S. (1992) The thermogenic role of exercise in the treatment of morbid obesity: a critical evaluation. *American Journal of Clinical Nutrition,* **55,** 533S–537S

Cannon, C. E. (1981) The diet of eastern and pale headed Rosellas. *Emu,* **81,** 101–110.

Chaudhuri, C. R. & Chatterjee, I. B. (1969) L-ascorbic acid synthesis in birds: phylogenetic trend. *Science,* **164,** 435–436.

Cunha, T. J. (1980) *Horse Feeding and Nutrition.* Academic Press: New York.

Davison, K. E., Potter, G. D., Greene, L. W., Evans, J. W. & McMullan, W. C. (1991) Lactation and reproductive performance of mares fed added dietary fat during late gestation and early lactation. *Journal of Equine Veterinary Science,* **11** (2), 111.

Deschutter, A. & Leeson, S. (1986) Feather growth and Development. *Worlds Poultry Science,* **42–43,** 259–267

DLG—Food values for horses. 2. Aufl. (1992) DLG-Verlag: Frankfurt/Main.

Dolensek, E. & Bruning, D. (1978) Ratites . In *Zoo and Wild Animal Medicine.* ed. M. E. Fowler. 1st edn. pp. 165–180. W. B. Saunders: Philadelphia.

Donoley, R. (1992) Zinc toxicity in caged and aviary birds — 'New Wire Disease'. *Australian Veterinary Practice,* **22,** 6–11.

Drepper, K., Menke, K. H., Schulz, G. & Wachter-Vormann, W. (1988) Investigations on the protein and energy requirements of adult budgerigars (*Melopsittacus undulatus*) in cage husbandry. *Kleintierpraxis,* **33,** 57–62.

Drochner, W. & Meyer, H. (1991) Digestion of organic matter in the large intestine of ruminants, horses, pigs and dogs. *Journal of Animal Physiology and Animal Nutrition,* **65,** 18–40.

Earle, K. E. & Clarke, N. R. (1991) The nutrition of the budgerigar (*Melopsittacus undulatus*). *Journal of Nutrition,* **121,** S186–S192.

Earle, K. E. & Smith, P. M. (1991a) Digestible energy requirements of adult cats at maintenance. *Journal of Nutrition,* **121** S45–S46.

Earle, K. E. and Smith, P. M. (1991b) The effect of dietary taurine content on the plasma taurine concentration of the cat. *British Journal of Nutrition,* **66,** 227–235.

Ellis, M. (1984) Canary, Bengalese and zebra finch. In *Evolution of Domesticated Animals.* ed. I. L. Mason. London: Longman.

Evans, H. E. (1969) Anatomy of the budgerigar. In *Diseases of Cage and Aviary Birds.* ed. M. L. Petrak. pp. 45–112. Philadelphia: Lea & Febiger.

Evans, M. (1992) UK's new vets at a disadvantage without knowledge of bird care. *Cage and Aviary Birds,* 1st February.

Featherstone, W. R. (1976) Glycine–serine interrelations in the chick. *Federation Proceedings,* **35,** 1910–1913

Finke, M. D. (1991) Evaluation of the energy requirements of adult kennel dogs. *Journal of Nutrition,* **121,** S22–S28.

Frape, D. (1986) *Equine Nutrition and Feeding.* Longman Group UK Limited.

Freeman, D. W., Potter, G. D., Schelling, G. T. & Kreider (1988) Nitrogen metabolism in mature horses at varying levels of work. *Journal of Animal Science,* **66,** 407.

Fridhandler, L. & Quaster, J. H. (1955) *Archives of Biochemistry and Biophysics,* **56,** 412.

Gesellschaft für Ernvöhrungsphysiologie der Haustiere (1982) Studies on the energy and nutrition of horses. DLG-Verlag: Frankfurt/Main.

Grau, C. R. & Roudybush, T. E. (1986) Lysine requirement of cockatiel chicks. *American Federation of Aviculture,* Research Report Dec–Jan, pp. 12–14.

Gray, G. M. (1975) *New England Journal of Medicine,* **292,** 1225.

Gray, G. M. & Santiago, N. S. (1966) *Gastroenterology*, **51**, 489.

Gray, G. M., Lally, B. C. & Conklin, K. A. (1979) *Journal of Biological Chemistry*, **248**, 25.

Gray, G. M., Walter, W. M. & Colver, E. H. (1968) *Gastroenterology,* **54**, 552.

Gries, C. L. & Scott, M. L. (1972a) The pathology of thiamin, riboflavin, pantothenic acid and niacin deficiencies in the chick. *Journal of Nutrition*, **102**, 1269–1285.

Gries, C. L. & Scott, M. L. (1972b) The pathology of pyridoxine deficiency in chicks. *Journal of Nutrition*, **102**, 1259–1267.

Griewe, K. M., Meacham, T. N., Fregin, G. F. and Walberg, J. L. (1989) Effect of added dietary fat on exercising horses. *Proceedings of the 11th Equine Nutrition and Physiology Society Symposium*, p. 101.

Griffiths, C. R. (1968) The geriatric cat. *Journal of Small Animal Practice*, **9**, 343–355.

Griminger, P. (1983) Digestive system and nutrition. In *Physiology and Behaviour of the Pigeon*. ed. M. Abs. London: Academic Press.

Guthrie, H. A. (1975) *Introductory Nutrition*. 3rd edn. St Louis, Missouri: C V Mosby Co.

Halver, J. E. (ed.) (1989) *Fish Nutrition*. 2nd edn. 798 pp. London: Academic Press.

Hambleton, P. L., Slade, L. M., Hamar, D. W., Kienholz, E. W. & Lewis, L. D. (1980) Dietary fat and exercise conditioning effect on metabolic parameters in the horse. *Journal of Animal Science*, **51**, 1330.

Harrison, G. J. & Harrison, L. R. (1986) Nutritional diseases. In *Clinical Avian Medicine and Surgery*. eds. G. J. Harrison & L. R. Harrison. Philadelphia: W. B. Saunders.

Hasholt, J. & Petrak, M. (1983) Gout. In *Diseases of Cage and Aviary Birds*. 2nd edn. pp. 639–645. Philadelphia: Lea and Febiger.

Hayes, K. C., Carey, R. E. & Schmidt, S. Y. (1975) Retinal degeneration associated with taurine deficiency in the cat. *Science*, **188**, 949–951.

Hazewinkel, H. A. W., How, K. L., Bosch, R., Goedegebuure, S. A. & Voorhout, G. (1990) Inadequate photosynthesis of vitamin D in dogs. In *Nutrition, Malnutrition and Dietetus in the Dog and Cat*. eds. H. Meyer, E. Kienzle, & A. T. B. Edney. pp. 66–68. London: British Veterinary Association.

Hegde, S. N. (1973) Composition of pigeon milk and its effect on growth in chicks *Indian Journal of Experimental Biology* , **11**, 238–239

Hintz, H. F. (1983) *Horse Nutrition—A Practical Guide*. New York: Arco Publications.

Hintz, H. F., Ross, M. W., Lesser, F. R., Leids, P. F., White, K. K., Short, J. E. & Schryver, H. F. (1978) The value of dietary fat for working horses I. Biochemical and haematological evaluations. *Journal of Equine Medicine and Surgery*, **2**, 483.

HMSO (1991) The feeding stuffs regulations 199. statutory instrument No. 2840. London: Her Majesty's Stationery Office.

Honigman, H. (1936) Studies on nutrition in mammals, Part I. *Proceedings of the Zoological Society of London*, 517– 530

Hopfer, U. (1987) Membrane transport mechanisms for hexoses and amino acids in the small intestine. In *Physiology of the Gastrointestinal Tract*. 2nd edn. Vol. 2. pp. 1499–1526. New York: Raven Press.

Kamphues, J. & Meyer, H. (1991) Basic data for factorial derivation of energy and nutrient requirements of growing canaries. *Journal of Nutrition*, **121**, S207–S208.

Kear, J. (1962) Food selection in finches with special reference to interspecific differences. *Proceedings of the Zoological Society of London*, **138**, 163–204.

Kendall, P. T., Holme, D. W. & Smith, P. M. (1982) Comparative evaluation of net digestive and absorptive efficiency in dogs and cats fed a variety of contrasting diet types. *Journal of Small Animal Practice*, **23**, 577–587.

Kienzle, E. (1988) Investigations on intestinal and intermediary metabolism of carbohydrates (starch of different origin and treatment, mono- and disaccharide) in the domestic cat (*Felis catus*). Hannover, Tierarztliche Hochschule, Habilschrift, 1989.

Kienzle, E. & Meyer, H. (1989) The effects of carbohydrate-free diets containing different levels of protein on reproduction in the bitch. In *Nutrition of the Dog and Cat*. eds. I. H. Burger & J. P. W. Rivers. pp. 243–257. Cambridge: Cambridge University Press.

Kienzle, E. & Rainbird, A. (1991) Maintenance energy requirements of dogs: what is the correct value for the calculation of metabolic body weight in dogs? *Journal of Nutrition*, **121**, S39–S40.

Kienzle, E., Radicke, S., Wilke, S., Landes, E. & Meyer, H. (1992) Pre-ileal starch digestion in relation to source and preparation of starch. *1st European Conference on the Nutrition of Horses*. Pferdeheilkunde (Sonderausgabe) S. 103–106.

Kim, Y. S. & Erickson, R. I. T. (1985) Role of peptidases of the human small intestine in protein digestion. *Gastroenterology*, **88**, 1017–1013.

Kimmich, G. A. & Randles, J. (1980) Evidence for an intestinal Na⁺–sugar transport coupling stoichiometry. *Biochimica et Biophysica Acta*, **596**, 439–444.

Kronfeld, D. S. (1982) Feeding dogs for hard work and stress. In *Dog and Cat Nutrition*. pp. 61–73. Ed. A. T. B. Edney. Oxford: Pergamon Press.

Kronfeld, D. S., Donoghue, S. & Glickman, L. T. (1991) Body condition and energy intakes of dogs in a referral teaching hospital. *Journal of Nutrition*, **121**, S157–S158.

Kronfeld, D. S., Hammel, E. P., Ramberg Jr, C. F. & Dunlap Jr, H. L. (1977) Haematological and metabolic responses to training in racing sled dogs fed diets containing medium, low or zero carbohydrate. *American Journal of Clinical Nutrition*, **30**, 419–430.

Lasiewski, R. C. & Dawson, W. R. (1967) A re-examination of the relation between standard metabolic rate and bodyweight in birds. *Condor*, **69**, 13–23.

Lawrence, L. A., Pagan, J., Pubols, M., Reeves, J., White, K., Douglas, R. & Gaskins, C. (1989) Influence of isocaloric high energy carbohydrate and fat diets on growth-related hormone profiles in the yearling horse. *Proceedings of the 11th Equine Nutrition and Physiology Society Symposium*, p. 151.

Legrand-Defretin, V (1993) Energy requirements of dogs and cats—what goes wrong? *International Journal of Obesity* (in press).

Lever, C. (1984) Budgerigar. In *Evolution of Domesticated Animals*, Chapter 54. ed. I. L. Mason. London: Longman.

Li, T. K. & Valle, B. L. (1980) The biochemical and nutritional role of other trace elements. In *Modern Nutrition in Health and Disease*. eds. R. S. Goodhardt & M. E. Shils. 6th edn. pp. 408–441. Philadelphia: Lea & Febiger.

Long, J. L. (1984) The diets of three species of parrots in the south of Western Australia. *Australian Wildlife Research*, **11** (2), 357–372.

Loveridge, G. G. (1986) Bodyweight changes and energy intakes of cats during gestation and lactation. *Animal Technology*, **37**, 7–15

Loveridge, G. G. (1987) Some factors affecting kitten growth. *Animal Technology*, **38**, 9–18

Loveridge, G. G. & Rivers, J. P. W. (1989) Bodyweight changes and energy intakes of cats during pregnancy and lactation. In *Nutrition of the Dog and Cat*. eds. I. H. Burger & J. P. W. Rivers. pp. 112. Cambridge: Cambridge University Press.

Lowenstine, L. J.(1986) Nutritional disorders of birds. In *Zoo and Wild Animal Medicine*. ed. M. E. Fowler. 2nd edn. pp. 201–212. Philadelphia: W. B. Saunders.

MacDonald, M. L., Anderson, B. C., Rogers, Q. R., Buffington, C. A. & Morris, J. G. (1984a) Essential fatty acid requirements of cats: pathology of essential fatty acid deficiency. *American Journal of Veterinary Research*, **45**, 1310–1317.

MacDonald, M. L., Rogers, Q. R. & Morris, J. G. (1984b) Nutrition of the domestic cat, a mammalian carnivore. *Annual Review of Nutrition*, **4**, 521–562.

Männer, K (1991) Energy requirement for maintenance of adult dogs. *Journal of Nutrition*, **121**, S37–S38.

Markham, R. W. & Hodgkins, E. M. (1989) Geriatric nutrition. *Veterinary Clinics of North America: Small Animal Practice*, **19**, 165–185

Markwell, P. & Gaskell, C. J. (1991) Progress in understanding feline lower urinary tract disease. *Waltham International Focus*, **1** (3), 22–29.

Massey, D. M., Sellwood, E. H. B. & Waterhouse, C. E. (1960) The amino acid composition of budgerigar diet, tissues and carcass. *Veterinary Record*, **72**, 283–286.

McCance R. A. & Widdowson E. M. (1991) *The Composition of Foods*. 5th edn. The Royal Society of Chemistry, Cambridge.

McDonald, P. Edwards, R. A. & Greenhalgh, J. F. D. (1988) *Animal Nutrition*. 4th edn. Harlow: Longman.

Meyer, H. (1987) *Nutrition of the equine athlete. Equine Exercise Physiology II*. eds J. R. Gillespie & W. E. Robinson. pp. 650–657. Davis: California.

Meyer, H. (1990) *Ernährung des Hurdes* (Nutrition of Dogs). 2nd edn. Stuttgart: Ulmer GmbH & Co.

Meyer, H. (1992) *Pferdefütterung* (2. Aufl.). Verlag Paul Parey, Berlin und Hamburg.

Meyer, H. & Kienzle, E. (1991) Dietry protein and carbohydrates: Relationship to clinical disease. *Proceedings Purina International Symposium* (in association with Eastern States Veterinary Conference, 1991) pp. 13–26.

Mooney, C. T. (1991) Problems of the ageing cat. *The Bulletin of the Feline Advisory Bureau*, **28** (3), 64–67

Morris, D. (1961) Seed preferences of certain finches under controlled conditions. *Avian Magazine*, **61**, 271–287.

Morris, J. G. & Rogers, Q. R. (1978) Arginine: an essential amino acid for the cat. *Journal of Nutrition*, **108**, 1944–1953.

Morton, S. R. & Davies, P. H. (1983) Food of the zebra finch (*Poephila guttata*) and an examination of granivory in birds of the Australian arid zone. *Australian Journal of Ecology*, **8**, 235–43.

Munday, H. S. & Davidson, H. P. B. Normal Gestation Lengths in the Domestic Shorthair Cat (*Felis domesticus*). Submitted for publication.

Munday, H. S. & Earle, K. E. (1991) The energy requirements of the queen during lactation and kittens from birth to 12 weeks. *Journal of Nutrition*, **121**, 543–44

National Research Council (1981) *Nutrient Requirements of Coldwater Fishes*. Washington D.C.: National Academy Press.

National Research Council (1983) *Nutrient Requirements of Warmwater Fishes and Shellfishes*. Washington D.C.: National Academy Press.

National Research Council (1984) *Nutrient Requirements of Poultry*. 8th edn. Washington D.C.: National Academy Press.

National Research Council (1985) *Nutrient Requirements of Dogs*. Washington D.C.: National Academy Press.

National Research Council (1986) *Nutrient Requirements of Cats*. Washington D.C.: National Academy Press.

National Research Council (1987) *Vitamin Tolerance of Animals*. Washington D.C.: National Academy Press.

National Research Council (1989) Nutrient Requirements of Horses, 5th edn. Washington D.C.: National Academy Press.

Nickel, R., Schummer, A. & Seiferle, E. (1979) *The Viscera of the Domestic Mammals*. 2nd edn. New York: Springer-Verlag.

Nott, H. M. R. (1992) The nutritional requirements of psittacines. *Waltham International Focus*, **2 (3)**, pp. 2–7

O'Neill, S. J. B., Blanave, D. & Jackson, N. (1971) The influence off eathering and environmental temperature on the heat production and efficiency of utilization of metabolizable energy by the mature cockerel. *Journal of Agricultural Science*, **77**, pp. 293–305.

Oldham, S. L., Potter, G. D., Evans, J. W., Smith, S. B., Taylor, T. S. & Barnes, W. S. (1990) Storage and mobilisation of muscle glycogen in exercising horses fed a fat-supplemented diet. *Journal of Equine Veterinary Science*, **10** (5), 53.

Pagan, J. D. & Hintz, H. F. (1986) Equine energetics I. Relationship between bodyweight and energy requirements in horses. *Journal of Animal Science*, **63**, 815–821.

Pagan, J. D., Tiegs, W., Jackson, S. G. & Murphy, H. Q, (1993) The effect of different fat sources on exercise performance in thoroughbred racehorses. *Proceedings 13th Equine Nutrition and Physiology Society Symposium, p. 125.*

Perry, R. A. (1983) *Diseases of Birds: Avian Therapy and Disease Control.* The T G Hungerford vade Mecum Series for Domestic Animals, No. 2. The University of Sydney post-graduate foundation in Veterinary Science.

Peterson, C. F., Lampman, C. E. & Stamberg, O. E. (1949) Effect of riboflavin on hatchability of eggs from battery confined hens. *Poultry Science*, **26**, 187–191.

Peterson, M. E. & Graves, T. K. (1992) Diagnosis and treatment of occult hyperthyroidism in cats. In *Proceedings of the 15th Waltham/OSU Symposium for the Treatment of Small Animal Diseases: Endocrinology.* eds. J. H. Sokolowski & W. W. Campfield. pp. 7–12. Vernon, California: Kal Kan Foods Inc.

Petrak, M. L. (1982) *Diseases of Cage and Aviary Birds*, 2nd edn. Philadelphia: Les & Febiger.

PFMA (1993) PFMA Profile. London: Pet Food Manufacturers' Association.

Pion, P. D., Kittleson, M. D. & Rogers, Q. R. (1989) Cardiomyopathy in cats and its relation to taurine deficiency. In *Current Veterinary Therapy, Vol. 10.* ed. R. W. Kirk. pp. 251–262. Philadelphia: W. B. Saunders.

Pion, P. D., Kittleson, M. D., Rogers, Q. R. & Morris, J. G. (1987) Myocardial failure in cats associated with low plasma taurine: a reversible cardiomyopathy. *Science*, **237**, 764–768.

Pitts, C. (1983) Hypovitaminosis A in Psittacines. In *Current Veterinary Therapy VIII.* ed. R. W. Kirk. Philadelphia: W. B. Saunders.

Potter, G. D., Gibbs, P. G., Haley, R. G. & Klendschoj, C. (1992) Digestion of protein in the small and large intestines of equines fed mixed diets. 1st European Conference on the Nutrition of Horses. *Pferde-heilkunde* (Sonderausgabe), S. 140–143.

Price, C. J. (1988) Nutritional diseases. In *BSAVA Manual of Parrots, Budgerigars and Other Psittacine Birds.* ed. C. J. Price. London: British Veterinary Association.

Rabinowitch, V. (1969) The role of experience in the development and retention of seed preferences in zebra finches. *Behaviour*, **4**, 33, 222–236.

Rainbird, A. & Kienzle, E. (1990) Studies on the energy requirement of dogs depending on breed and age. *Kleintierpraxis*, **35**, 149–158.

Randell, M. G. (1981) Nutritionally induced hypocalcemic tetany in an Amazon parrot. *JAVMA*, **179**, 1277–1278.

Rivers, J. P. W. (1982) Essential fatty acids in cats. *Journal of Small Animal Practice*, **23**, 563–576.

Rivers, J. P. W., Frankel, T. L., Juttla, S. & Hay, A. W. M. (1979) Vitamin D in the nutrition of the cat. *Proceedings of the Nutrition Society*, **38**, 36A.

Robben, J. H. & Lumeij, J. T. (1989) Comparative studies on parrot foods commercially available in the Netherlands. *Tijdscrift voor diergeneeskunde*, **114** (1), 19–25.

Roberts, M. C. (1975) Carbohydrate digestion and absorption studies in the horse. *Research Veterinary Science*, **18**, 64

Rogers, Q. R. & Morris, J. G. (1982a) Do cats really need more protein? *Journal of Small Animal Practice*, **23**, 521–532.

Rogers, Q. R. & Morris, J. G. (1982b) *3rd Annual Pet Food Institute Technical Symposium.* Washington D.C.: Pet Food Institute.

Romsos, D. R., Palmer, H. J., Muiruri, K. L., Bennink, M. R. (1981) Influence of a low carbohydrate diet on performance of pregnant and lactating dogs. *Journal of Nutrition*, **111**, 678–689.

Rosskopf, W. J. & Woerpel, R. W. (1991) Pet avian conditions and syndromes of the most frequently presented species seen in practice. *Veterinary Clinics of North America, Small Animal Practice*, **21** (6), 1189–1211.

Roudybush, T. E. & Grau, C. R. (1986) Food and water interrelations and the protein requirement for growth of an altricial bird, the cockatiel (*Nymphicus hollandicus*). *Journal of Nutrition*, **116**, 552–559.

Roudybush, T. E. & Grau, C. R. (1991) Cockatiel (*Nymphicus hollandicus*) nutrition (abstract). *Journal of Nutrition*, **121**, S206.

Ryan, T. (1991) Trace elements and their role in avian nutrition. *Canine Practice*, **16** (2), 30–35.

Saunders, D. A. (1980) Food and movements of the short-billed form of the white-tailed black cockatoo. *Australian Wildlife Research*, **7**, 257–269.

Schaefer, A. E., Salmon, W. D. & Strength, D. R. (1949) Inter-relationship of vitamin B$_{12}$ and choline. *Proceedings of the Society of Experimental Biology and Medicine*, **71**, 202–204.

Scott, B. D., Potter, G. D., Evans, J. W., Reagor, J. C., Webb, G. W. & Webb, S. P. (1989) Growth and feed utilisation by yearling horses fed added dietary fat. *Journal of Equine Veterinary Science*, **9** (4), 210.

Seawright, A. A. & English, P. B. (1967) Hypervitaminosis A and deforming cervical spondylosis of the cat. *Journal of Comparative Pathology*, **77**, 29–39.

Sheffy, B. E., Williams, A. J., Zimmer, J. F. & Ryan, G. D. (1985) Nutrition and metabolism of the geriatric dog. *Cornell Veterinary*, **75**, 324–347.

Slater, P. J. B. (1974) The temporal pattern of feeding in the zebra finch. *Animal Behaviour*, **22**, 506–515.

Snyder, N. F. R., King, W. B. & Kepler, C. B. (1982) Biology and conservation of the Bahama Parrot. *Living Bird*, **19**, 91–114.

Snyder, R. L. & Terry, J. (1986) Avian nutrition. In *Zoo and Wild Animal Medicine*, M. E. Fowler. 2nd edn. Philadelphia: W. B. Saunders.

Steffens, W. (1989) *Principles of Fish Nutrition* 384 pp. Chichester UK: Ellis Horwood Ltd.

Sturkie, P. D. (1954) *Avian Physiology*, Comstock Publishing Associates

Sturman, J. A., Gargano, A. D., Messing, J. M. & Imaki, H. (1986) Feline maternal taurine deficiency: effect on mother and offspring. *Journal of Nutrition*, **116**, 655–667.

Taylor, S. (1954) Calcium as a goitrogen. *Journal of Clinical Endocrinology and Metabolism*, **14**, 1412–1422.

Thornburg, L. P., Dennis, G. L., Olwin, D. B., McLaughlin, C. D. & Gulbas, N. K. (1985b) Copper toxicosis in dogs, Part 2: the pathogenesis of copper-associated liver disease in dogs. *Canine Practice*, **12** (5), 33–38.

Thornburg, L. P., Ebinger, W. L., McAllister, D. & Hoekema, D. J. (1985a) Copper toxicosis in dogs,

Part I: copper-associated liver disease in Bedlington Terriers. *Canine Practice*, **12** (4), 41–45.

Ullrey, D. E., Allen, M. E. & Baer, D. J. (1991) Formulated diets versus seed mixtures for psittacines. *Journal of Nutrition*, **121**, S193–S205.

Van den Broek, A. H. M. & Thoday, K. L. (1986) Skin diseases in dogs associated with zinc deficiency: a report of five cases. *Journal of Small Animal Practice*, **27**, 313–323.

Wallach, J. D. (1970) Nutritional diseases of exotic animals. *Journal of American Veterinary Medical Association*, **157**, 583–599.

Wallach, J. D. & Flieg, G. M. (1967) Nutritional secondary hyperthyroidism in captive psittacine birds. *Journal of American Veterinary Medical Association*, **151**, 880–883.

Wallach, J. D. & Flieg, G. M. (1969) Nutritional secondary hyperthyroidism in captive birds. *Journal of American Veterinary Medical Association*, **155**, 1046–1051.

Wannemacher, R. W. & McCoy, Jr (1966) Determination of optimal dietary protein requirements of young and old dogs. *Journal of Nutrition*, **88**, 66–74.

Warner, A. C. I. (1981) Rate of passage of digesta through the gut of mammals and birds. *Nutrition Abstracts Reviews*, **51**, 789–820.

Watkins, B. A. (1991) Importance of essential fatty acids and their derivatives in poultry. *Journal of Nutrition*, **121**, 1475–1485.

Windmueller, H. G. (1982) Glutamine utilization by the small intestine. *Advances in Enzymology*, **53**, 201–237.

Wolter, R., Boidot, J. P. & Morice, M. (1970) Essais de détermination des besoins azotés du Pigeon de rapport. *Record Médicine Veterinary*, **146**(1), 1–13.

Wood, H. O. (1944) The surface area of the intestinal mucosa in the rat and cat. *Journal of Anatomy*, **78**, 103–105.

Wyndham, E. (1980a) Diurnal cycle, behaviour and social organisation of the budgerigar *Melopsittacus undulatus*. *Emu*, **80**, 25–33

Wyndham, E. (1980a) Environment and food of the budgerigar *Melopsittacus undulatus*. *Australian Journal of Ecology*, **5**, 47–61.

Wysocki, A. A. & Baker, J. P. (1975) Utilisation of bacterial protein from the lower gut of the equine. *Proceedings of the 4th Equine Nutrition and Physiology Symposium*, p. 21–43.

Zann, R. & Straw, B. (1984) Feeding ecology and breeding of zebra finches in farmland in Northern Victoria. *Australian Wildlife Research*, **11**, 533–52

Zazula, M. (1984) Changes in body protein level in tree sparrow (*Passer montanus*) induced by high and low protein diets. *Ekologia Polska*, **32**, 709–720

Zeuner, F. E. (1963) *A History of Domesticated Animals*. London: Hutchinson.

Zmija, G (1991) Feeding procedures in galloping and trotting horses. *Veterinary Dissertation*, Hanover Vet. School.

Appendices

NATIONAL RESEARCH COUNCIL NUTRIENT REQUIREMENTS OF GROWING DOGS AND CATS			
Nutrient	**Units**	**Dog**[a]	**Cat**[b]
Protein	g	NS	11.4
Arginine	mg	327	478
Histidine	mg	117	144
Isoleucine	mg	234	239
Leucine	mg	380	574
Lysine	mg	335	383
Methionine and cystine	mg	253	359
Phenylalanine and tyrosine	mg	466	407
Threonine	mg	304	335
Tryptophan	mg	98	72
Valine	mg	251	287
Taurine	mg	NR	19
Fat	g	3.2	NS
Linoleic acid	g	0.64	0.24
Arachidonic acid	mg	NR	9.53
Minerals			
Calcium	mg	382	382
Phosphorus	mg	287	287
Potassium	mg	287	191
Sodium	mg	36	24
Chloride	mg	55	91
Magnesium	mg	26	19
Iron	mg	2.1	3.8
Zinc	mg	2.3	2.4
Copper	µg	191	239
Manganese	µg	335	239
Iodine	µg	38	17
Selenium	µg	7.2	4.8
Vitamins			
Vitamin A (retinol)	IU	240	158
Vitamin D (cholecalciferol)	IU	26.4	24
Vitamin E (α-tocopherol)	IU	1.5	1.4
Vitamin K (phylloquinone)	µg	NS	4.7
Thiamin	µg	65	239
Riboflavin	µg	163	191
Pantothenic acid	µg	645	239
Niacin	µg	717	1912
Pyridoxine	µg	71.7	191
Folic acid	µg	12.9	38.2
Vitamin B_{12}	µg	1.7	0.96
Choline	mg	81	115
Biotin	µg	NS	3.3

[a]National Research Council, 1985, average 3 kg Beagle puppy requiring 2.5 MJ/d.
[b]National Research Council, 1986.
NS Not stated.
NR Not required.

APPENDIX Ia: National Research Council nutrient requirements of growing dogs and cats per MJ of metabolisable energy.

ESSENTIAL AMINO ACID REQUIREMENTS FOR MAINTENANCE		
Amino acid	**Dog[a]**	**Cat**
Arginine	68	478[b]
Histidine	71	ND
Isoleucine	155	ND
Leucine	271	ND
Lysine	161	158[c]
Methionine and cystine	97	155[c]
Phenylalanine and tyrosine	277	ND
Threonine	142	ND
Tryptophan	42	ND
Valine	193	ND

[a]Data from NRC (1985).
[b]Data from NRC (1986) for the near-adult cat.
[c]Data from Burger and Smith (1990).
ND No research data.

APPENDIX Ib: Essential amino acid requirements for adult maintenance: mg per MJ metabolisable energy.

WCPN MINIMUM NUTRIENT REQUIREMENTS FOR DOGS AND CATS							
Nutrient	Units	Adult maintenance		Growth		Reproduction	
		Dog	Cat	Dog	Cat	Dog	Cat
Protein	g	9.6	15	13	17	13	17
Fat	g	3.3	5	3.3	5	3.3	5
Linoleic acid	g	0.66	0.6	0.66	0.6	0.66	0.6
Arachidonic acid	mg	NR	10	NR	12	NR	12
Minerals							
Calcium	g	0.39	0.39	0.39	0.39	0.66	0.60
Phosphorus	g	0.30	0.30	0.30	0.30	0.53	0.48
Ca/P ratio		0.5	0.5	0.8	0.8	0.8	0.8
Sodium	g	0.04	0.02	0.05	0.12	0.05	0.12
Potassium	g	0.3	0.3	0.3	0.3	0.3	0.3
Magnesium	mg	23	18	23	18	23	30
Iron	mg	2.4	3.8	4.8	3.8	4.8	6.0
Copper	mg	0.3	0.24	0.3	0.24	0.42	0.3
Manganese	mg	0.3	0.24	0.3	0.24	0.3	0.6
Zinc	mg	3	2.4	3	2.4	3	2.4
Iodine	mg	0.04	0.02	0.04	0.02	0.09	0.06
Selenium	mg	6	4.8	6	4.8	6	6.0
Vitamins							
Vitamin A	IU	245	159	245	159	299	329
Vitamin D	IU	26	24	26	24	30	60
Vitamin E	mg	1.8	1.4	3	1.4	3	4.8
Vitamin K	µg	a	6.0	a	6.0	a	6.0
Thiamin	mg	0.06	0.24	0.06	0.24	0.06	0.3
Riboflavin	mg	0.15	0.19	0.15	0.19	0.15	0.3
Pantothenic acid	mg	0.66	0.24	0.66	0.24	0.66	0.6
Niacin	mg	0.72	1.9	0.72	1.9	0.72	2.7
Pyridoxine	mg	0.07	0.19	0.07	0.19	0.07	0.24
Folic acid	µg	13	38	13	38	13	60
Vitamin B_{12}	µg	1.6	0.96	1.6	0.96	1.6	1.2
Choline	mg	75	119	75	119	75	119
Biotin	µg	a	4.2	a	4.2	a	4.2
Taurine — canned food	mg	NR	149	NR	149	NR	149
Taurine — dry food	mg	NR	60	NR	60	NR	60

NR No requirement.
a No requirement when natural ingredients are fed. This is because intestinal bacterial synthesis can generally meet the needs of the animal.

APPENDIX Ic: WCPN minimum nutrient requirements for dogs and cats per MJ metabolisable energy.

WCPN NUTRIENT REQUIREMENTS OF ORNAMENTAL FISH			
Nutrient	**Unit**	**Range**	**Notes**
Protein: digestible energy ratio	g/MJ	37.0 50.0	Dependent on life-stage and species
Protein	% of diet	30.0 50.0	Dependent on life-stage and species
Essential amino acids			Dependent on life-stage and species
Threonine	% of protein	2.2 4.0	
Valine	% of protein	3.0 4.0	
Isoleucine	% of protein	2.2 4.0	
Leucine	% of protein	3.3 5.3	
Phenylalanine	% of protein	5.0 6.5	with 0.0% tyrosine
Phenylalanine	% of protein	3.1 4.1	with 0.4% tyrosine
Lysine	% of protein	5.0 5.7	
Histidine	% of protein	1.5 2.1	
Methionine	% of protein	2.3 4.0	with 0.0% cysteine
Methionine	% of protein	1.5 2.4	with 1.0% cysteine
Arginine	% of protein	4.3 6.0	
Essential fatty acids			
Linoleic acid	% of diet	1.0	w3:w6 ratio dependent on salinity and temperature
Linolenic acid	% of diet	1.0	optimum of species
Vitamins			Dependent on life-species and species
Vitamin A	IU/kg	5000 20000	
Vitamin D$_3$	IU/kg	2000 4000	
Vitamin E	mg/kg	100 500	
Vitamin K$_3$	mg/kg	10 20	
Vitamin B$_1$	mg/kg	10 20	
Vitamin B$_2$	mg/kg	10 20	
Niacin	mg/kg	50 150	
Pyridoxine	mg/kg	10 20	
Vitamin B$_{12}$	mg/kg	0.02 0.05	
Pantothenic acid	mg/kg	50 250	
Folic acid	mg/kg	5 10	
Inositol	mg/kg	300 500	
Choline chloride	mg/kg	1000 2000	
Biotin	mg/kg	1.0 1.5	
Vitamin C	mg/kg	200 400	
Minerals			
Calcium	mg/kg	300 7000	
Phosphorus	mg/kg	4000 6000	
Calcium: phosphorus	mg/kg	1:1 1:3	
Potassium	mg/kg	6000 12000	
Sodium chloride	mg/kg	12000 30000	
Magnesium	mg/kg	400 700	Dependent on mineral content of water
Iron	mg/kg	200	Dependent on life-stage and species
Zinc	mg/kg	80 200	
Manganese	mg/kg	12 13	
Copper	mg/kg	3 5	
Selenium	mg/kg	0.5–1.0	
Cobalt	mg/kg	0.05	
Fluorine	mg/kg	1.00	

APPENDIX Id: WCPN nutrient requirements of ornamental fish. All figures are ranges (not minima and maxima) and are dependent on species and life stage as well as diet composition.

DAILY NUTRIENT REQUIREMENTS FOR WORKING HORSES						
Nutrient	**Light**		**Moderate**		**Intensive**	
	<300 (kg BW)	300–800 (kg BW)	<300 (g BW)	300–800 (kg BW)	<300 (g BW)	300–800 (kg BW)
DE[a]	20	14.2	24	16.8	26	18.2
DP[b]	95	67.5	120	84	130	100
	per 100 kg BW		per 100 kg BW		per 100 kg BW	
Calcium (g)	5.2		5.3		6	
Phosphorus (g)	3		3		3.1	
Magnesium (g)	1.6		1.7		2.1	
Sodium (g)	4.7		7.2		19	
Potassium (g)	6.5		8		15	
Chloride (g)	12		16.2		36	
Iron (mg)	100–120					
Copper (mg)	10–15					
Zinc (mg)	100–120					
Manganese (mg)	80–100					
Cobalt (µg)	200–250					
Selenium (µg)	300–350					
Iodine (µg)	300					
Vitamin A (IU)	7500–8000					
Vitamin D (IU)	500–1000					
Vitamin E (IU)	200–250					
Thiamin (mg)	2–3					
Riboflavin (mg)	1.2–1.5					
Biotin (mg)	1–2					

[a] Digest. Energy (MJ/100 kg BW).
[b] Digest. Crude Protein (g/100 kg BW).
From Meyer (1992), DLG (1992) and NRC (1989).

APPENDIX Ie: Daily nutrient requirements for working horses.

DAILY NUTRIENT REQUIREMENTS FOR FOALS					
Nutrient	**6–12 months**		**13–24 months**		**25–36 months**	
	<300 (kg BW)	300–800 (kg BW)	<300 (g BW)	300–800 (kg BW)	<300 (g BW)	300–800 (kg BW)
DE[a]	16–20	12–13	16–20	13.5–14.5	17–21	14.5–15.5
DP[b]	120–140	90–95	100–120	85–90	95–115	75–85
	per 100 kg BW		per 100 kg BW		per 100 kg BW	
Calcium (g)	5.8		5.5		5.1	
Phosphorus (g)	3.9		3.8		3.2	
Magnesium (g)	1.0		1.4		1.6	
Sodium (g)	1.4		1.7		2.1	
Potassium (g)	3.2		4.2		5.1	
Chloride (g)	4.4		6.3		8.0	
Iron (mg)	180–250					
Copper (mg)	25–30					
Zinc (mg)	120–150					
Manganese (mg)	100–120					
Cobalt (µg)	250–300					
Selenium (µg)	350–400					
Iodine (µg)	500–600					
Vitamin A (IU)	20000–25000					
Vitamin D (IU)	1500–2500					
Vitamin E (IU)	150–200					
Thiamin (mg)	2–3					
Riboflavin (mg)	1.2–1.5					
Biotin (mg)	1–2					

[a] Digest. Energy (MJ/100 kg BW).
[b] Digest. Crude Protein (g/100 kg BW).
From Meyer (1992), DLG (1992) and NRC (1989).

APPENDIX If: Daily nutrient requirements for foals.

NUTRIENT REQUIREMENTS FOR HORSES – MAINTENANCE/GESTATION/LACTATION						
Nutrient	**Maintenance**		**Gestation 9–11 months**		**Lactation 1–3 months**	
	<300 (kg BW)	300–800 (kg BW)	<300 (g BW)	300–800 (kg BW)	<300 (g BW)	300–800 (kg BW)
DE[a]	18	13.5	23	17.5	30–38	24–27
DP[b]	95	66	120–150	100–110	250–330	210–235
	per 100 kg BW		per 100 kg BW		per 100 kg BW	
Calcium (g)	5.0		8.0		12	
Phosphorus (g)	3.0		5.5		9	7.5
Magnesium (g)	1.5		1.6		2.3	2.1
Sodium (g)	2.5		2.5		3.0	
Potassium (g)	5.0		5.3		8.0	
Chloride (g)	8.0		8.0		9.6	
Iron (mg)	100–120		180–200			
Copper (mg)	10–15		24–26			
Zinc (mg)	100–120		120–140			
Manganese (mg)	80–100		100–110			
Cobalt (µg)	200–250		250–270			
Selenium (µg)	300–350		350–370			
Iodine (µg)	300		500–550			
Vitamin A (IU)	7500–8000		10000–15000			
Vitamin D (IU)	500–1000		1500–1700			
Vitamin E (IU)	150–200		150–200			
Thiamin (mg)	2–3		2–3			
Riboflavin (mg)	1.2–1.5		1.2–1.5			
Biotin (mg)	1–2/day		2–3/day			

[a] Digest. Energy (MJ/100 kg BW).
[b] Digest. Crude Protein (g/100 kg BW).

APPENDIX Ig: Daily nutrient requirements for maintenance, gestation and lactation.

ENERGY REQUIREMENTS: DOGS AND CATS								

Energy content of commercial dog and cat foods: recommended factors for conversion of nutrient content to energy are as follows (per g):

	Dog (from NRC, 1985)		Cat (from NRC, 1986)					
			Canned		Semi-moist		Dry	
	kcal	kJ	kcal	kJ	kcal	kJ	kcal	kJ
Protein	3.5	14.6	3.9	16.3	3.7	15.5	5.65	23.6
Fat	8.5	35.6	7.7	32.2	8.8	36.8	9.4	39.3
[a]Carbohydrate	3.5	14.6	3.0	12.6	3.3	13.8	4.15	17.4
[b]Constant			-0.05	-0.21			-1.26	-5.27

(Protein/Fat/Carbohydrate for Dry $\times$ 0.99)

Energy requirements of animals (WCPN recommendations)

Dog	Cat (per kg bodyweight per day)		
		kcal	kJ
Adult maintenance (M)[c] =	Adult maintenance (normal)	70–90	293–377
125 $W^{0.75}$ kcal/day	Adult maintenance (inactive)	50–70	209–293
523 $W^{0.75}$ kJ/day	Pregnancy	100–140	419–586
Highly active adult: up to 2M	Lactation	240+	1004+
Lactation: up to 3M	Growth: 10 weeks old	220	921
Pregnancy (from week 6): 1.3M	20 weeks old	160	670
Growth: Young: 2M	30 weeks old	120	502
Half-grown: 1.5M	40 weeks old	100	419
Near-adult: 1.2M			

[a] The carbohydrate value is usually calculated "by difference" i.e. it is the residual fraction after protein, fat, moisture and ash have been analysed. For this reason carbohydrate by difference is often also referred to as nitrogen-free extract (NFE).
[b] The constants are calculated on the basis of kcal/g of product.
[c] Assumes a moderate amount of activity; W is body weight in kg.
Note: $W^{0.75}$ can be easily computed using a basic calculator, as it is the fourth root (i.e. square root twice) of the cube value of the body weight.

APPENDIX II: Energy requirements of dogs and cats.

Europe

In Europe pet foods are controlled principally by lesgislation whcih covers animal feeding stuffs including feeds for farm animals. Since farm animals form part of the human food chain, the laws are necessarily stringent. The regulations originate primarily in European Community (EC) directives which are then implemented through national regulations. For example, in the United Kingdom these directives are promulgated through the Feeding Stuffs Regulations (HMSO, 1991). The industry is also governed by codes of practice set by industry trade associations such as the UK Pet Food Manufacturers' Association (PFMA, 1993).

The principal EC directives involved in the legislation of pet foods are as follows:

• **Marketing of compound feeding stuffs** — concerned with the labelling and distribution of compound feeding stuffs. A compound feeding stuff is a mixture of products of vegetable or animal origin and therefore most, if not all, pet foods will fall under this heading (Council Directive 79/373/EEC).

• **Additives in feeding stuffs** — concerned with the use of additives in all feeds including prepared pet foods (Council Directive 70/524/EEC).

• **Undesirable substances and products** — concerned with setting maximum permitted levels for undesirable substances (e.g. heavy metals) in feeding stuffs, including pet foods (Council Directive 74/63/EEC).

• **Certain products** — concerned with products which act as direct or indirect protein sources in feeding stuffs including pet foods (Council Directive 82/471/EEC).

• **Categories of ingredients** — concerned with the grouping of ingredients into various categories e.g. meat and animal derivaties, fish and fish derivatives (Commission Directive 82/475/EEC).

• **Assessment** — there are also directives concerned with setting guidelines for the assessment of additives (Council Directive 87/153/EEC) and certain products (Council Directive 83/228/EEC) used in animal nutrition.

USA

In the United States of America, pet food is regulated at both the federal and state levels. On the federal level, the applicable legislation is the Federal Food, Drug and Cosmetic Act (FD&C Act) which regulates the composition and labelling of "food" in interstate commerce in the USA. Section 201(f) of the FD&C Act defines "food" as including food used for man or other animals e.g. pet or companion animals.

Pet food is also regulated by the individual states of the USA. The "Uniform State Feed Bill" (and regulations) adopted by the Association of American Feed Control Officials (AAFCO) includes pet food within its definition of "commercial feed". In its "Official Publication" — also known as the "AAFCO Manual" — AAFCO provides definitions for various categories of animal feed and pet food ingredients (AAFCO, 1993).

Most states have adopted a version of the United States Feed Bill which permits the states to regulate pet food as commercial feed. In addition, most states have also adopted a version of the Official Pet Food Regualtions approved by AAFCO in conjunction with the pet food industry's representatives operating through the Petfood Institute (PFI). The US federal and state regulation of pet food is discussed further below.

Federal

Under the FD&C Act, pet food is regulated by the United States Food and Drug Administration (FDA). The statute prohibits the distribution of "adulterated" or "misbranded" pet food.

With regard to additives and ingredients permitted in pet food, FDA has promulgated regulations authorising permitted additives (21 Code of Federal Regulations (CFR) Part 573), and regulation identifying those additives or ingredients that may be used because they are "generally recognised as safe" (GRAS) (21 CFR Part 584). FDA has also promulgated regulations for substances which are prohibited from use in animal or pet feed (21CFR Part 589). Although additives and ingredients may not be identified in FDA regulations for use in animal or pet food, as long as such additives or ingredients is generally permitted in the USA for pet food. FDA has also promulgated specific package labelling requirements for animal feed, including pet food (21 CFR Part 501). These regulations are "mirrored" substantially by the labelling provisions in the Uniform State Feed Bill and Official Pet Food Regulations.

Through its Centre for Veterinary Medicine (CVM), FDA has begun to take a more active role in pet food safety and some labelling issues. For example, in recent years, FDA has been reviewing certain pet food labels to determine whether there is a scientific basis to approve certain types of health-related claims on pet food labels. It is expected that FDA's role in this area will continue in conjunction with the Pet Food Committee of AAFCO.

continued

APPENDIX III: Legislation.

LEGISLATION (continued)

State

As indicated previously, virtually all US states have commercial feed legislation — generally a version of the AAFCO Uniform State Feed Bill — which serves to regulate pet food in the individual states. In addition, most states have a version of the Pet Food Regulations as well. It is generally these state requirements which serve as the primary basis for regulation of pet food in the USA.

Via state Commercial Feed legislation, virtually all pet food product labels marketed in the USA must be "approved" or registered" in each state. Interestingly, there is no similar requirement for most human food products in the USA, and thus, commercial feed and pet food can be viewed as being subject to more stringent regulation than human food.

As indicated above, ingredients or additives used in pet food should comply either with an applicable FDA regulation or AAFCO manual feed ingredients definition.

The Pet Food Regulations govern definitions and terms (PF-1), label format and labelling (PF-2), brand and product names (PF-3), expression of guarantees (PF-4), ingredients (PF-5), directions for use (PF-6) and drugs and pet food additives (PF-7). As a general proposition, compliance with these regulations is normally required by individual states before pet food products and labels will be registered.

Composition of packaging and pet food processing

The composition of food packaging must comply with FDA requirements. FDA has detailed regulations for food packaging which would include the materials used in the packaging of pet food. Pet food packaging must be evaluated to assure compliance with these regulations.

FDA also has regulations for thermally processed and aseptically packaged pet foods. For example, for canned pet foods, the manufacturing plant must register with FDA and the process or "cook" must also be filed.

Imported pet food products

Prior to importation of pet food into the USA, customs regulations and tariffs must be checked. In addition, the US Department of Agriculture, Animal and Plant Health Inspection Service (APHIS) has prohibitions and/or criteria involving the importation of various ingredients (e.g. ruminant-derived materials and fowl) into the USA. APHIS regulation found in 21 CFR Parts 94 and 95 should be reviewed before considering importation of pet food products or ingredients into the US.

As far as the pet owner is concerned, the main outward expression of this legislation is the pack label. As indicated above, there are strict regulations which require manufacturers to list various characteristics of the product. This declaration goes under various descriptions, but in the UK is known as the statutory statement and in the USA as the principal display and information panels. The requirements vary to some extent from one country to another but generally include an ingredient list, a guaranteed analysis, a statement on nutritional adequacy and directions for use (Burger and Thompson, 1993).

From AAFCO (1993); Burger and Thompson (1993); HMSO (1991); PFMA (1993).

BIOLOGICAL TRIAL PROCEDURES

In Chapter 1 the importance of assessing the nutritional performance of a food was discussed. The process by which a biological trial is designed, conducted and evaluated is summarised below.

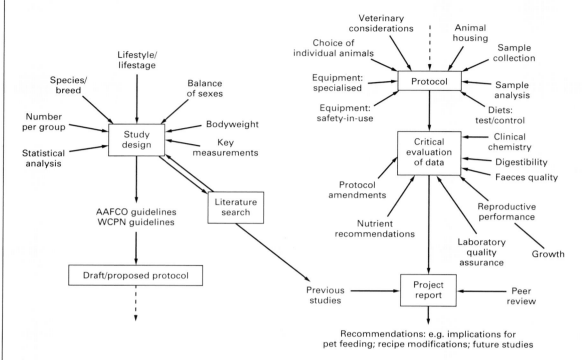

For a typical pet food, the assessment usually starts with a comparison of the nutrient content of the product of the product with an appropriate standard recommendation (see Appendix I). Ultimately however, the food must be tested in the animal for which it is intended: *in vivo* assessment. There are essentially four principal testing procedures:

> • Digestibility
> • Growth
> • Reproduction
> • Adult maintenance

The first stage is usually the digestibility study which is designed to measure the availability of the nutrient content (see Chapter 2, section on energy). The average daily amount of a nutrient absorbed by the animal is calculated from the difference between the nutrient intake in food and output in faeces. Digestibility is usually expressed as a percentage thus:

$$\text{Digestibility \%} = \frac{\text{Intake} - \text{faecal output}}{\text{Intake}} \times 100$$

Since faeces do not consist only of undigested, unabsorbed material but contain cell debris and material excreted into the digestive tract, the difference between intake and output measured in this way is defined as apparent digestibility or apparent absorption. To measure true absorption or true digestibility, it is necessary to use control diets free of the nutrient being studied to establish the size and output when intake is zero. For most practical purposes apparent digestibility is the measurement used as it measures the net amount of digestion. Within a species, digestibility values are largely independent of the individual animal and are more a characteristic of the food.

Growth trials measure the most rapid period of development which in the case of dogs and cats is between about 6 and 18 weeks of age. For reproduction the assessment starts at the commencement of pregnancy (gestation) and ends when the young are weaned. Maintenance studies for dogs and cats usually last a minimum of 6 months. A comprehensive description of these studies can be found in AAFCO (1993).

The assessment of adequacy is either in relation to a control diet tested at the same time and which is known to give satisfactory results, or by comparison with the historical average performance for the colony of animals where the test is being conducted.

APPENDIX IV: Biological trial procedures.

Index